Chapter I

Introduction

Astronomy is said to be the oldest and noblest of the physical sciences. Yet, the earliest astronomers were astrologers who studied the movement of the heavenly bodies to determine their influences on the lives of men and nations. Because of the clear atmosphere prevailing or the plains and plateaus of Assyria, Babylonia, Persia, Egypt, Arabia, China, India and Greece, ancient astrologers were able to read the position of the heavenly bodies with the naked eye. As civilization moved westward to regions where the atmosphere was denser it became more difficult to read the heavens with the naked eye. The invention of the telescope solved this difficulty, but the cost of this instrument put it beyond the reach of most astrologers, hence they became more and more dependent on tables known as Ephemerides, prepared by their more fortunate brethren who had access to telescopes.

Thus, the fundamentals of astronomy, which were a matter of daily observation to the early astrologers, became a lost art to modern astrologers. Most astrological textbooks devote little or no space to this subject, which has long been looked upon as an occult art requiring years of study, a thorough knowledge of higher mathematics, and a deep understanding of fundamental science. Actually, this is not the case. Recent progress in simplified methods of instruction and solutions of problems has reduced the mathematics involved to simple addition and subtraction. A simple knowledge of the apparent and real motions of the heavenly bodies can thus be acquired in a few hours of study.

This book is based on, and adapted from, the text so successfully used by the U.S. Naval Academy, the U.S. Maritime Academy, and the flying schools of the U.S. Air Force and of the commercial airlines, in instructing the hundreds of thousands of young men, who with little or no knowledge of mathematics become navigators of ships and planes during World War II. Just as the navigator determines his position at sea or in the sky by reference to his angular relationship to two or more known heavenly bodies at a given instant of time, so does the astrologer chart the angular distances between the heavenly bodies at a given place and instant of time. The astrologer's chart of the heavens is known as the horoscope, and may be erected for the time of birth (natal astrology), the time of a query (horary astrology) or the time of an election (electional astrology).

Geocentric Vs. Heliocentric Astronomy

In primitive times, communication was slow and difficult; hence each race lived unto itself and had very little contact with other races. Each race naturally felt that it was living at the center of the Earth's surface, and certain places were considered to be the centers of all astronomical calculations, thus Babylon for the Assyrians, Mt. Olympus for the Greeks, Jerusalem for the Jews, etc. In more modern times, Greenwich, England, was gradually accepted as the international reference point for astronomical calculations. In recent years, Washington, D.C. is being accepted by American astrologers as the reference point.

It was likewise only natural for the ancients to consider the Earth as the center of the universe, for the heavenly bodies seemed to revolve around the Earth. Although some early Greek philosophers like Pythagoras (6th century B.C.) taught that the Earth revolved around the Sun, which was the center of the universe, their views carried little weight, and the ideas of the great astronomer Ptolemy, who lived in Alexandria, Egypt about A.D. 140 were taught for about 1,500 years. Not until A.D. 1543 did the Polish astronomer, Copernicus, dare propose the theory that the Earth revolved on its axis, thus accounting for the rising and setting of the heavenly bodies, and that the Earth was only one of a number of heavenly bodies that moved around the Sun in circles. The Copernican theory was corrected and revised by Galileo in Italy, Kepler in Germany, and Newton in England, and is today the accepted system of astronomy. Ptolemy's system, which considered the Earth as the center of the universe is known as the Geocentric, and the Copernican system, which has the Sun as the center of the universe, is known as the Heliocentric System. See Figures 1 and 2.

However, since the navigator wants to know where he is with reference to the Earth, and the astrologer wants to know the position of the heavenly bodies in relation to a particular place on the Earth, modern astronomers have provided tables showing the position of those bodies as viewed from the Ephemeris and Nautical Almanac are prepared by the Nautical Almanac Office of the United States Naval Observatory and are issued under an international agreement held in October 1911 at Paris, France. The labor of preparing these tables is shared with offices of the British Nautical Almanac, the Berliner Jahbuch, and the Connaissance des Temps, and the International Astronomical Union. Astrologers then convert these tables into the form shown in the Ephemerides commonly used.

Simplified Astronomy for Astrologers

LCdr. David Williams

Copyright 2009 by American Federation of Astrologers

No part of this book may be reproduced or transcribed in any form or by any means, electronic or mechanical, including photocopying or recording or by any information storage and retrieval system without written permission from the author and publisher, except in the case of brief quotations embodied in critical reviews and articles. Requests and inquiries may be mailed to: American Federation of Astrologers, Inc., 6535 S. Rural Road, Tempe, AZ 85283.

ISBN-10: 0-86690-172-8
ISBN-13: 978-0-86690-172-7

Cover Design: Jack Cipolla

Published by:
American Federation of Astrologers, Inc.
6535 S. Rural Road
Tempe, AZ 85283

www.astrologers.com

Printed in the United States of America

Contents

Chapter 1, Introduction	1
Chapter 2, Coordinating Time and Space	17
Chapter 3, The Age of ARies the Warrior	31
Chapter 4, The Age of Pisces the Mystic	41
Chapter 5, The Aquarian Age	59
Chapter 6, The Solar System	73
Chapter 7, Time	83
Chapter 8, The Calendar and Ancient Chronology	91
Chapter 9, Eclipses	99
Chapter 10, Evolution of the Planisphere and House Division	109

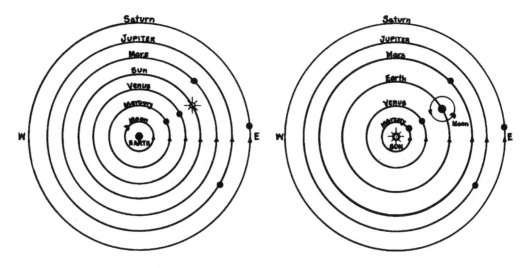

Fig. 1, Geocentric System of Ptolemy *Fig. 2, Heliocentric System of Copernicus*

Systems of Coordinates

(a) **Terrestrial Coordinates.** The geographer locates specific places on Earth by expressing their distances north or south of the equator in circles parallel to the equator and to each other, called parallels of latitude. These are numbered from 0° at the equator to 90° at the poles. Circles running through the north and south poles and crossing the parallels of latitude at right angles are called meridians of longitude. The meridian passing through Greenwich, England was chosen by international agreement as the starting point and is called the prime meridian. The meridians are numbered from 0° to 180° east or west of Greenwich. These relationships are shown in Figure 3.

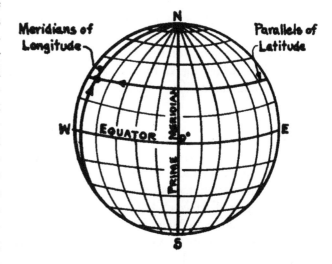

Fig. 3, Terrestrial Coordinates Latitude-Longitude

(b) **Celestial Coordinates.** The astronomer employs exactly similar methods in expressing the location of heavenly bodies on the surface of the celestial sphere. But instead of one system of coordinates, he uses several, namely, the ecliptic, equator, and horizon systems, each designed for a different purpose.

(1) Ecliptic System. The system used by the ancient astronomers is known as the ecliptic system, since they confined their studies largely to the moving bodies, such as the Sun, Moon and planets, which appeared to travel in a narrow 18° band called the zodiac, which was centered on the ecliptic, the apparent path of the Sun in the heavens.

For all practical purposes, the heavens may be considered to form a huge dome or sphere of infinite radius called the Celestial Sphere, with the Earth at its center. The Earth's axis extended cuts the celestial sphere in two points called the north and south celestial poles. It is around this axis that the celestial sphere appears to rotate. The plane of the Earth's equator extended till it meets the celestial sphere is called the Celestial Equator or Equinoctial. Since the Earth rotates on its axis from west to east, the Sun appears to move about the Earth from east to west in a great circle in the celestial sphere called the ecliptic, which cuts the celestial equator at two points called the equinoctial points, at an angle of about 23°27′, called the obliquity of the ecliptic.

The equinoctial point occupied by the center of the Sun on or about March 21, at the instant it moves north of the equator is called the Vernal Equinox, or the first point of Aries, which marks the commencement of spring in the northern hemisphere. The equinoctial point occupied by the center of the Sun on or about September 21, at the instant is moves south of the equator, is call the Autumnal Equinox, or the first point of Libra, which marks the beginning of autumn in the northern hemisphere.

In the ecliptic system, the angular distance north or south of the ecliptic is known as celestial latitude, but it is not to be confused with terrestrial latitude. The angular distance east of the vernal equinox measured along the ecliptic is called celestial longitude, but is likewise is not to be confused with terrestrial longitude. Most ephemerides used by astrologers give the positions of the heavenly bodies in terms of celestial longitude and latitude. These relationships are shown in Figure 4.

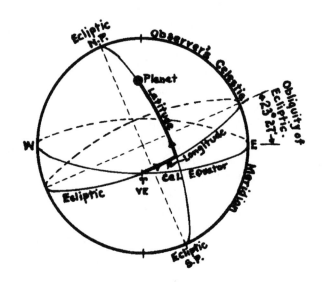

Fig. 4, Ecliptic System Coordinate Celestial Latitude-Longitude

(2) **Equator System.** Because of the obliquity of the ecliptic, we have seen that the coordinates of the ecliptic system do not correspond with those of Earth. Hence, modern astronomers have devised the equator system by means of which any point on Earth may be projected onto the celestial sphere, or vice versa. The coordinates of this system can be measured with greater precision than those of any of the others, and it is customary to obtain the others from this system by trigonometric calculations.

In this system, the planes of Earth's equator and the parallels of latitude, extended till they cut the celestial sphere, form the celestial equator or equinoctial and the parallels of declination. Similarly, the Earth's meridians of longitude, extended till they cut the celestial sphere, form the celestial meridians.

Just as the geographer measures distances along the Earth's equator from a starting point (the Greenwich Meridian) so does the astronomer measure distances along the celestial equator from a starting point—the vernal equinox or first point of Aries. Distance measured along the celestial equator eastward of the Vernal Equinox to the hour circle of a heavenly body (which is a great circle passing thru the celestial poles and the heavenly body), is called the right ascension of the body and is expressed in hours, from 0^h to 24^h R.A. Right ascension is thus similar to longitude on Earth. Distances north or south of the celestial equator are measured on the hour circle passing through the heavenly body and are expressed as declination, from $0°$ at the celestial equator to $90°$ at the celestial poles. Declination is thus similar to latitude on Earth. Right ascension and declination of celestial bodies are used to express their positions in space and their positions relative to each other.

To express the position of a heavenly body in relation to the Earth's meridians of longitude, the term hour angle is used. The hour angle of a celestial body is the distance between the celestial meridian of the observer (which is his longitude), and the hour circle passing thru the body. It is called the local hour angle (LHA) of the body and is always measured westward from the meridian, from $0°$ to $360°$ (sometimes from $0°$ to $24°$). The local hour angle measured from the celestial meridian of Greenwich is called the Greenwich hour angle (GHA). Since the celestial meridian of the observer remains stationary with respect to a place on the Earth (the observer's longitude), while the hour circle moves with the body as it circles the Earth from east to west, hour angles always increase westward. GHA and LHA always differ by the longitude of the observer, or LHA—GHA—Longitude W or GHA/Longitude E.

When the hour angle is measured either eastward or westward of the celestial meridian from 0 to 180 E or W (or 0^h to 12^h) it is called the meridian angle (t) of the celestial body and is the value used in the solution of the astronomical triangle. "t" is found from LHA. If LHA is greater than 180, t = 360 – LHA and is labeled E. The sidereal hour angle (SHA) of a celestial body is the distance between the hour circle of the Vernal Equinox and the Vernal Equinox thru $360°$. While both RA and SHA measure the angular distance between the same hour circles, they differ only in (1) direction of measurement and (2) the units of measurement. The position of the Fixed Stars is expressed in SHA which is equal to 360 – the RA of the star in degrees, or SHA = 360 – RA (in degrees). All of the foregoing relationships are shown in Figure 5.

(3) **Horizon System.** We have seen that by means of the equator system of coordinates, namely the celestial body's Greenwich hour angle and declination, the body position may

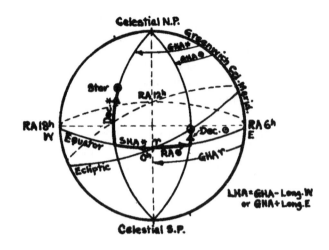

Fig. 5, Equator System Coordinates Declination-Right Ascension

be plotted on Earth's surface as longitude and latitude. This position is known as the geographical position (GP) or the subsolar, sublunar, or substellar point of the celestial body. To an observer at this geographic position, the body would be in his zenith, or that point of the celestial sphere vertically overhead. Thus, if the declination of a body were 26N and the GHA were 115; an observer at that instant of time would have the body in his zenith if he were located at latitude 26N and longitude 115W.

But the navigator wants to know where he is at a particular instant of time by measuring his angular relationship to two or more celestial bodies; and the astrologer wants to plot the position of all the planets, as well as the cusps of houses at a specified instant of time and in relation to a particular place on Earth. To do this a third system of coordinates based on the position of the observer, called the horizon system, must be used.

If an observer looks about him, he would see a circle where Earth appears to meet the sky in what is called the visible horizon. But as the size of this circle varies with the height of the observer's eye, a plane at right angles to a line connecting the observer's zenith and nadir, and passing thru the center of the Earth to intersect the celestial sphere in a great circle called the celestial horizon, is chosen as the plane of reference. A vertical circle passing thru the zenith and nadir of the observer and the north and south points of the horizon is called the meridian of the observer. Another vertical circle passing thru the zenith and nadir of the observer and the east and west points of the horizon is called the prime vertical. The north point of the horizon is that point of intersection of the meridian and horizon that is nearest the North Pole.

Fig. 6, Horizon System Coordinates Altitude-Azimuth

The angular distance from the celestial horizon to a heavenly body measured on a vertical circle is called the altitude (H) of the body. The angular distance between the meridian of the observer and the vertical circle passing thru the body is called azimuth angle (Z) of the body, and may be measured from the north or south point to the right, clockwise; or left, counterclockwise, thru 90° or 180°. It must be labeled N or S as a prefix and the direction of measurement E or W as a suffix thus S80W or N80E, etc. It is found by solution of the navigational triangle. A third term used in navigation is azimuth (Z) or true bearing, and it is the distance from the North Point of the horizon to the vertical circle of the body measured clockwise from the North Point thru 360°. Altitude and azimuth constitute the co-ordinates of the Horizon System and are shown in Figure 6.

Since meridian angle (t) and declination (d) are the coordinates of the equator system and altitude (H) and azimuth (Z) are the coordinates of the horizon system, the solution of a celestial observation is essentially a problem of converting the coordinates of one system to those of the other. This is accomplished by solving the astronomical triangle, which is a spherical triangle on the celestial sphere formed by the elevated pole, the zenith of the assumed position, the body being observed, and arcs of great circles connecting these points. In practice, the navigator determines the altitude (H) of a celestial body by sextant observation. For the exact instant of observation, determined by the use of a chronometer, he then computes the meridian angle t of the body using an assumed longitude. He obtains the declination (d) of the body at his instant from an ephemeris. Thus, knowing the values for meridian angle (t) and declination (d) he assumes a latitude (L) and solves the astronomical triangle for the body's computed altitude (H) corresponding to his assumed position. The actual computation is effected by tables and requires the use of no mathematics other than addition and subtraction. These tables are available online from the U.S. Naval Observatory.

It is interesting to note that modern celestial navigation is based upon a principle discovered by an American shipmaster, Capt. Thomas Hubbard Sumner, who, when lost in a storm off the

Irish coast on December 17, 1837, discovered that a single observation of the altitude of a heavenly body put the observer on a line of position that enables him to reach safety. Sumner's original method was later modified by Admiral Marcq Saint Hilaire of the French Navy, about 1870, and is the one now universally used.

At this point you may ask, of what value is the horizon system to the astrologer? To answer this question let us consider the following facts: The daily rotation of the Earth upon its axis from west to east causes the celestial bodies to appear to move across the sky from east to west, rising on the eastern point of the horizon, climbing the heavens

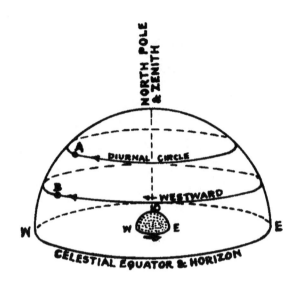

Fig. 7, The Parallel Sphere at Poles

until they reach the zenith of the observer's meridian, and declining there from to finally set at the western point of the horizon. Circles of the celestial sphere in which this daily movement appears to take place are called diurnal circles and coincide with the parallels of declination.

But the position of the diurnal circle of a body relative to the observer's horizon varies with the observer's latitude. In other words, each person carries his own zenith and horizon around with him, so to speak, so that the altitude and azimuth of a body as seen from New York, for instance, would be different from what they would be if the body was seen from New Orleans.

In Figure 7, the observer (O) is at the North Pole. Thus, his zenith and the North Pole coincide, and so do the celestial equator and celestial horizon. Let A and B be fixed stars. Their declinations being practically constant, each remains at the same angular distance from the equinoctial and therefore from the horizon, since in this case the horizon and the equinoctial coincide.

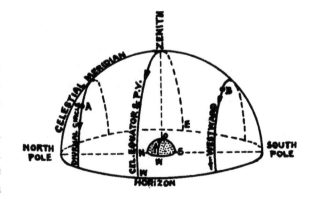

Fig. 8, The Right Sphere at Equator

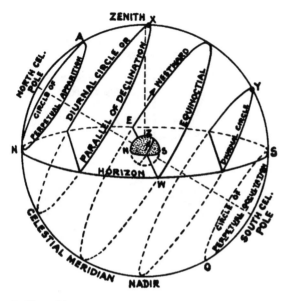

Fig. 9, The Oblique Sphere

Hence, the diurnal circle coincides with the declination circle and altitude circle. As Earth rotates to the eastward, the stars appear to revolve westward in diurnal circles, which are parallel to the equinoctial, and therefore, parallel to the horizon. Thus, they remain continuously above the horizon of the observer. When the Sun crosses the equinoctial into north declination on March 21, the Arctic regions have six months of daylight, while the south polar region has six months of darkness. On September 21 the reverse takes place. This aspect of the heavens is called the parallel sphere.

In Figure 8, the observer (O) is at the equator. The zenith lies in the equinoctial, which coincides with the prime vertical, and the poles lie in the horizon. The diurnal circles of the celestial bodies are perpendicular to the horizon and are half above and half below the horizon hence the bodies will only be visible half of the time. This aspect of the heavens is called the right sphere, and the term right ascension is derived there from, because for an observer at the equator the sidereal interval between the rising of the vernal equinox and the rising of a fixed star is equal to the right ascension of the star.

But in most cases, the observer is neither at the pole nor at the equator, but somewhere in between as in Figure 9, where the plane of the horizon is oblique to the planes of the equinoctial and the diurnal circles. If the observer is in north latitude, anybody (X) who is north of the equinoctial will be above the observer's horizon for more than 12 hours, while anybody (Y), who is south of the equinoctial, will be above the observer's horizon for less than 12 hours. Those bodies whose declination places them within or north of the circumpolar circle (AN) will not set. This circle is known as the circle of perpetual apparition. Similarly bodies within or south of the circle (SO) will never appear above the horizon. This circle is called the circle of perpetual occultation. The Southern Cross, for example, is never visible to most U.S. residents.

The three systems of coordinates with which we are most concerned, namely: the terrestrial, equator, and horizon systems are tied to each other thru the theorem: "The Latitude of a place is equal to the Altitude of the Elevated Pole, and is also equal to the declination of the Zenith." Thus, if the observer in Figure 9 is located in latitude 40N, the altitude of the North Star (which is close to the celestial North Pole) will always be 40° and every star within the Circle of Perpetual Apparition, or within 40° of the North Celestial Pole, will remain above the horizon all of the time, moving slowly around the North Star from east to west.

Precessional Path of the Celestial Poles

We are now ready to look into the phenomenon of precession. The Earth is an oblate spheroid, i.e., it bulges at the equator and is flattened at the poles. Its axis is inclined 23°27′ from the perpendicular to the plane of its orbit. If the Earth did not rotate on its axis, the gravitational pull of the Moon (and to a lesser extent of the Sun) on the Earth's bulging equator would bring the equator into the plane of the Moon orbit (on the average, the ecliptic plane). But the whirling motion of the Earth resists this tendency, and in accordance with the law of gyroscopic motion, the result is a slow conical movement of the earth's axis, westward around the vertical to the plane of the ecliptic. See Figure 10. This conical movement of the earth's axis results in a circular motion of the celestial pole westward around the ecliptic pole, called the precessional path of the celestial pole. The Earth completes one precessional cycle in 25,794 years for the Epoch of 1844, or the pole moves westward approximately 50″.2453 per year, according to Prof. Simon Newcomb's formula, which is based on values tabulated from A.D. 1600 to A.D. 2100. The value for the precessional cycle varies slightly, depending on the Epoch Year chosen as the starting point in applying Newcomb's Formula.

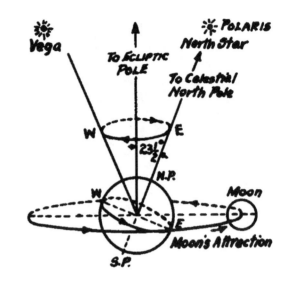

Fig. 10, Precession of Poles

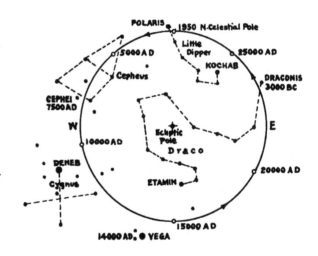

Fig. 11, Precessional Path of North Celestial Pole

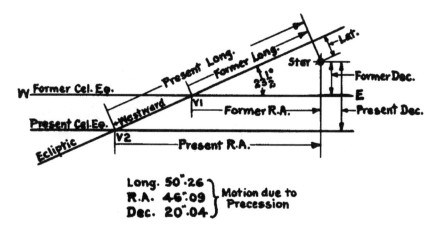

Fig. 12, Effect of Precession of Equinoxes

At the present time, the North Celestial Pole is 1° from Polaris and moving closer to it. Its nearest approach will be in about A.D. 2100 when it will be 28′ away. Then it will continue on along its precessional path, leaving Polaris behind, until in about A.D. 7500 alpha Cephei and in A.D. 14000. Vega (alpha Lyrae) becomes successively "pole stars." Similarly, alpha Draconis was the "pole star" about 3000 B.C. See Figure 11. We have seen that the circles of perpetual apparition, and occultation are at a distance from the poles equal to the observer's latitude, hence precession makes a very marked difference in the constellations that are visible at a given place. For example, in the year 3000 B.C. the Southern Cross, which is now visible only in the extreme southern part of the U.S., could be seen from as far north as the site of Quebec.

Precession of the Equinoxes

The westward circular movement of the celestial pole causes a similar westward circular movement of the equator along the ecliptic. As the equator is a great circle, the equinoxes, or points where it cuts the ecliptic, must remain opposite each other, and any motion of the equator necessitates a corresponding motion of the equinoxes. This westward movement of the equinoxes along the ecliptic is called the Precession of the Equinoxes. In Figure 12, we see that as the Vernal Equinox moves westward from V to V, the right ascension, declination and longitude of the star are increased. If the change in longitude is 50″.26, the change in right ascension is 46″.09 and in declination 20″.04. The declination of a star at the opposite point of the sphere would be diminished, but the other two coordinates would still be increased. Latitude, which is measured from the ecliptic, is unaffected by precession. The tropical year (year of the seasons) is shorter than the sidereal year (true period of the Earth's revolution), because the vernal equinox moves westward to meet the Sun in its annual circuit of the ecliptic. It amounts to a little more than twenty minutes (50″.26/360 X365.25 days = 20.4 min.).

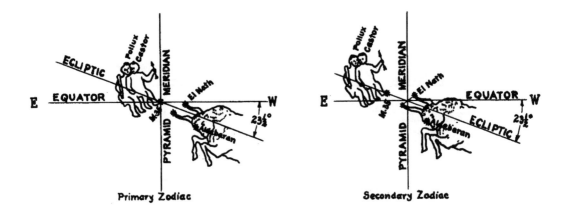

Fig. 13, Midnight Autumnal Equinox 4699 B.C.   Fig. 14, Midnight Autumnal Equinox 4000 B.C.

David Davidson, in his monumental book, *The Great Pyramid*, cites several recognized authorities who confirm his own independent calculations that originally the zodiac had six constellations of 60° each, starting with Taurus, the Bull. The first recorded zodiacal year started with the Full Moon at midnight of the Autumnal Equinox of 4699 B.C., at which time the meridian locating the future site of the Great Pyramid of Giza passed thru the star representing the toe of Castor, marking the beginning of the constellation of Gemini and the end of Taurus. See Figure 13. A secondary system of twelve constellations of 30° each was also in use, and at midnight of the autumnal equinox of September 22, 4000 B.C., the pyramid meridian cut in half the line joining the stars representing the tips of the Bull's horns, which were respectively above and below the ecliptic. See Figure 14. The Full Moon appeared as if balanced on the tips of the Bull's horns. Thus, in 699 years the Autumnal Equinox had regressed almost 10° of arc due to precession. Davidson's pyramid formula and Newcomb's formula show that 90° of precession from 4699 B.C. brings the Autumnal Equinox to A.D. 1843, thus in 6,541 years the Autumnal Equinox had regressed through 90° of the original zodiac due to the phenomenon of precession. Davidson gives the year A.D. 2500 as the date for 90° of precession from the secondary zodiac year of 4000 B.C. In both zodiacs, 90° of precession bring the Autumnal Equinox into the constellation of Pisces.

At about 1780 B.C., a compromise system of twelve irregular constellations was adopted, with the year beginning at the vernal equinox and the month beginning with the New Moon. Generally speaking, the constellations were alternately 40° and 20° wide. Thus, Taurus extended from the toe of Castor to the tail of Aries, a distance 40°, leaving only 20° for Aries, out of the original 60° extent of Taurus. Hence, the first point of Aries in the compromise system was the same as the first point of Taurus in the original system of six constellations. Midnight of the Autumnal Equinox in the original system and midnight of the Vernal Equinox in the compromise system

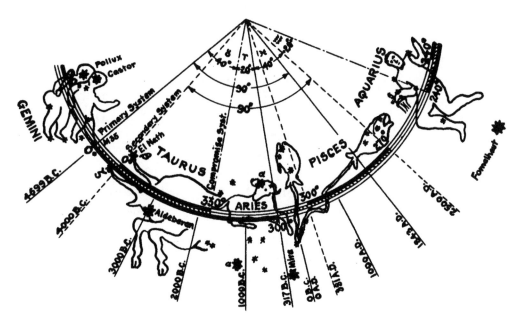

Fig. 15, Precession of the equinoxes through the three zodiac systems of antiquity. Adapted from The Great Pyramid by D. Davidson.

coincided in 317 B.C. From this point, the Vernal Equinox of 1950 will have regressed 31½° into the compromise constellation of Pisces. In the secondary system, the first point of Aries marked the equinox of A.D. 381. All three systems are shown in Figure 15.

The difficulty of determining when the so-called Aquarian Age is to commence can thus be seen to be due to the different zodiacal systems that have been used at various periods. If the original system were divided into twelve equal constellations of 30° each, beginning with the toe of Castor in 4699 B.C., 90° of precession through Taurus, Aries, and Pisces would indicate 1844 as the commencement of the Aquarian Age. In the secondary system, the Aquarian Age does not commence until A.D. 2500. Modern writers have added to the confusion by assuming different starting points for their zodiacal constellations, thus Paul Councel uses A.D. and Cyril Fagan, A.D. 213, for the first point of Aries. Astronomers make confusion worse confounded by calling the moving Vernal Equinox the first point of Aries and using an entirely different system of constellation sizes. The following tabulation shows the spread in the dates assigned by various authorities:

Authority	First Point of Aries	First Point of Aquarius
Cheiro	388 B.C.	1762 A.D.
D. Davidson (original and compromise system)	317 B.C.	1844 A.D.
A.M. Harding	300 B.C. (about)	?
Gerald Massey	255 B.C.	1905 A.D.
C.A. Jayne, Jr.	254 B.C.	1906 A.D.
Thierens	125 B.C.	?
Dane Rudhyar	97 B.C.	?
Paul Councel	0 A.D.	2160 A.D.
Cyril Fagan	13 A.D.	2369 A.D.
D. Davidson (Secondary System)	381 A.D.	2500 A.D.

Figure 16 portrays the zodiac at the Vernal Equinox of A.D. March 21, 1844. and clearly shows that as a result of 90° of precession from 4699 B.C., the Vernal Point has regressed from Star M35 in the toe of Castor to the tail of the Western Fish in Pisces. Thus, the toe of Castor now marks the Winter Solstice instead of the Vernal Equinox. The ecliptic has been divided into signs of 30° each, starting with the toe of Castor marking the 30th degree of Taurus or first degree of Gemini.

It should be understood, that for the purposes of astronomy, the constellations are regions of the celestial sphere set off by arbitrary boundaries, the last revision having occurred in 1928 when the International Astronomical Union made the boundaries parts of circles parallel and perpen-

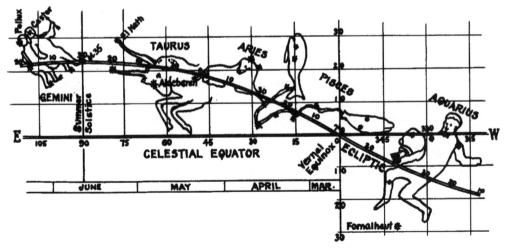

Fig. 16, Zodiac at Vernal Equinox of A.D. 1844

dicular to the celestial equator. The 48 original constellations described in Ptolemy's *Almagest* have been added to, and today 88 constellations are recognized. The constellations are useful for describing the approximate positions of the stars and other celestial bodies. Thus, the statement that Vega is in the constellation Lyra serves the same purpose as the information that Cleveland is in Ohio—we know about where it can be found.

The signs of the zodiac, however, were mapped out by the ancient priesthoods of Egypt and Mesopotamia in milestones corresponding to the twelve months of the year. Hence, the ecliptic circle, or apparent path of the Sun, is divided into 30 sections starting at the Vernal Equinox, where the Sun appeared to cross the equator on its northward journey. These sections were called signs, and were named after groups of stars whose rising and setting positions roughly corresponded to that of the Sun at a particular season. The names of these zodiac star clusters are: Aries, Taurus, Gemini, Cancer, Leo, Virgo, Libra, Scorpio, Sagittarius, Capricorn, Aquarius, and Pisces. The zodiac of the signs is sometimes called the moving zodiac, since the first point of Aries is continually moving westward because of the precession of the equinoxes.

Chapter 2

Coordinating Time and Space

When primitive man turned from the unsettled and precarious life of a hunter and began to lead a pastoral life around which an agrarian economy developed, he was confronted with the social necessity of keeping track of time and the season. It was essential that he know the right time to plant and to harvest; the right time for lambing and shearing his flocks; the time when the annual river floods would come, etc. Long before 4000 B.C. he had learned to count days by the rising and setting of the Sun, months by the changing phases of the Moon, and years by the heliacal rising of Sirius.

To recognize the direction in which an object will be seen on the horizon we must observe it from some fixed point, and be able to refer to some fixed line. Two fundamental lines of reference, the prime meridian which joins the north and south points on the horizon and the line at right angles to it joining the east and west points, had probably been settled before the great calendar civilizations began. The choice of the north and south meridian was based on the position of the Sun's shadow when it is shortest at midday. It also points directly toward the place in the heavens around which the stars seem to revolve at night. At the time when pyramids were built, alpha Draconis, a bright star in the constellation Draco, revolved in a tiny circle of three degrees around it.

The exact location of the noon shadow was determined by the use of man's first time and space measuring device, the gnomon or shadow-pole. This was at first merely a stick stuck in the sand or soft earth, and around which was traced a circle. The two points where the shadow touched the circle was bisected. A line connecting the point of bisection on the circle, with the shadow pole, pointed due north and south. A line at right angles to this, pointed due east and West. See Figure 17.

In the course of time, the simple shadow pole grew into elaborately carved obelisks, which were the "town clocks" of the ancient civilized world. Two of these obelisks, known as "Cleopatra's Needles" were originally erected by Thothmes III. One was sent to the Thames Embankment, London, 1878, and the other to Central Park, New York City in 1881, where they can be seen to this day. By means of obelisks such as these, the ancient Egyptian and Chaldean priests were able to determine the time of the equinoxes and the solstices (see Figure 18), as well as the obliquity of the ecliptic (see Figure 19). Altitude and azimuth were obtained as shown in Figure 20.

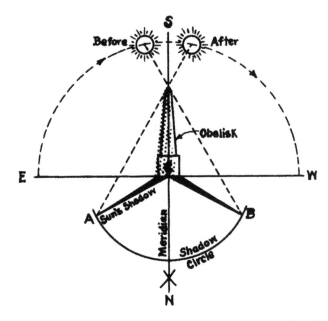

Fig. 17, Locating the meridian by bisecting the angle formed by the Sun's shadow just before and after noon

Perhaps the greatest astronomical observatory of ancient times was the Great Pyramid of Cheops at Giza, Egypt, in longitude 31E15, latitude 29N59. This structure was built about 2600 B.C. and is located at the exact center of the land area of the Earth. It was so oriented that the rays of Sirius at its transit of the meridian running thru the center of the pyramid, shone down the shaft leading from the south face to the king's chamber. The rays of alpha Draconis at its lower transit of this meridian shone down the two shafts on the north side. See Figure 21. A "scored line" in the entrance passage marks the date of the ob-

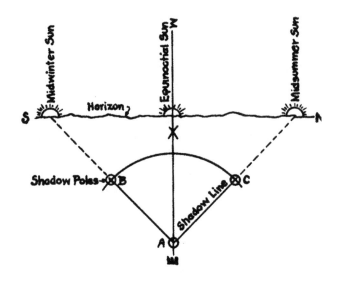

Fig. 18, Fixing the day of the equinoctial by bisecting the angle formed by the Sun's shadow at the solstices.

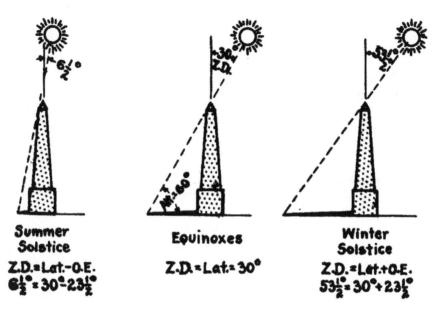

Fig. 19, Measuring obliquity of ecliptic from Sun's noon shadow

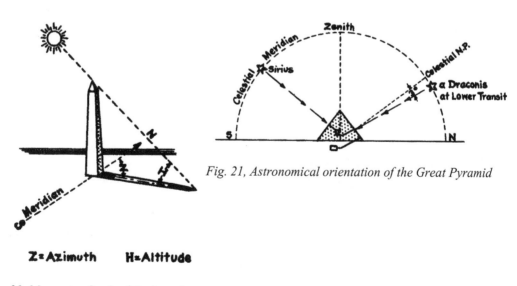

Fig. 21, Astronomical orientation of the Great Pyramid

Fig. 20, Measuring Sun's altitude and azimuth

served position of Alcyone of the Pleiades at midnight, Autumnal Equinox of 2144 B.C., according to Davidson, who states that Proctor's calculated date for Alcyone was 2140 B.C. and Brunnow's calculated date for alpha Draconis was 2136 B.C. See Figure 22.

The central location of the Great Pyramid and the structure itself are believed to have been foreordained as an imperishable marker or datum point from which astronomical and historical calculations could be made at some future time. The need for such an imperishable record has been illustrated at various times in the chequered history of mankind, first, when the Romans burned the classical library of antiquity at Alexandria in 40 B.C., and later when the riotous mobs repeated the destruction of the new Alexandrian library in A.D. 391. The Chinese have had their share of book burnings, and in modern times, the Nazis and Russians have vied with each other in an orgy of destruction of books that were in conflict with the peculiar ideologies of these enemies of civilization.

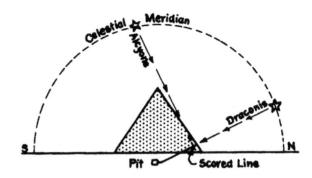

Fig. 22, Alcyone and Draconis at Great Pyramid in 2144 B.C.

Using Davidson's calculation of the time span covered by 90° of precession from 4699 B.C. as a basis, we divide the ecliptic circle into twelve equal segments of 30° each, beginning with Taurus. Reducing the ecliptic degrees to degrees of right ascension by means of DeLuce's *Table of Right Ascensions*, we superimpose the resulting divisions on a map of the world, with the pyramid meridian running thru the middle of Taurus. The result is shown in Figure 23, derived from the data shown in the tabulation beginning on page 14, which also contains a comparison with the conclusions of other writers.

The following explanation will be helpful to those who do not have a table of R.A.

Solution of Spherical Right Triangles
Since the ecliptic is inclined at an angle of 23°27' to the equator, an hour circle passing thru any point on the ecliptic forms a spherical right triangle, which may be represented thus:

Napier's "rules of circular parts" offer the easiest method of solving spherical right triangles. Eliminating the 90° angle, there remain five "parts," namely, the hypotenuse and two sides and the two remaining angles. For the purpose of Napier's rules, the five "parts" are the two sides, and the complements of the hypotenuse and the two angles arranged in the five sectors of a circle in the same order in which they occur in the triangle; thus:

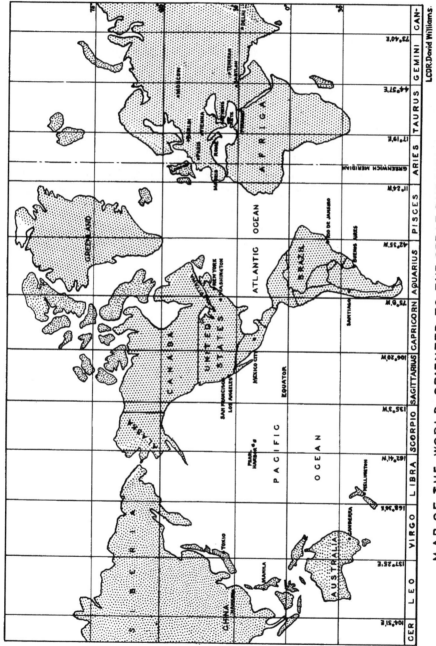

MAP OF THE WORLD ORIENTED TO THE GREAT PYRAMID OF GIZEH
DATUM—15° TAURUS = 31° 7′ 57″ E. LONG.
FIG. 23
LCDR. David Williams.

Napier's rules are: The sine of the middle part (which may be any one of the five parts) is equal to the product of:
(1) the tangents of the adjacent parts,
(2) the cosines of the opposite parts.

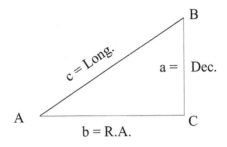

Suppose we know the ecliptic longitude and the obliquity of the ecliptic, we can find the right ascension as follow: Underscore the known parts c and A, then A, is a middle part with c and b as adjacent parts.
Thus, by Formula (1) sin Co-A = Co-C tan b, but since the function of an angle is equal to the co-function of the complement of the angle, this becomes (2) Cos A = cot c tan b. Multiplying both sides by 1 divided by Cot c and clearing, we get (3) Cos A times 1 divided by Cot c = tan b, but 1 divided by Cot c = tan c; therefore, by substitution (4) Cos A tan c = b or (5) O.E. tan long. = tan R.A. Example: Given long. 15°, O.E. 23°27′, find R.A. Using formula (5) and logarithms we get,

To log Cos 23°27′ = 9.96256
Add log tan 15°00′ = 9.42805
Ans log tan 13°49′ 19.39061-10

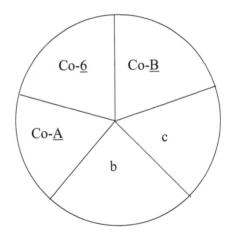

22

Table of Geodetic Equivalents of Zodiacal Divisions Based on Precession of Vernal Equinox from the Great Pyramid Meridian at 31°15' E Long., and 15° Taurus as Zero Datum Points.

			Terrestrial Longitude or Geodetic Equivalent Stated by				
Ecliptic Degree	Celes. Long	R.A.	Williams	Councel	Spencer	Jensen	Sepharial
Age of Taurus, the Bull, 4699 B.C. to 2508 B.C							
30° Taurus	345°	346°11'	45°04'E	47°E	45°40'E	61°07.5'E	57°48'E
15 Taurus	0	0 00	31 15 E	31 15 E	31 15 E	-	-
0 Taurus	15	13 49	17 26 E	17 E	18 11 E	31 07.5 E	27 54
Age of Aries, the Ram, 2508 B.C. to 317 B.C							
30 Aries	15°	13 49	17 26 E	17 E	18 11 E	31 07.5 E	27 54 E
0 Aries	45	42 32	11 17 W	11 W	11 49 W	1 07.5 E	0 00
Age of Pisces, the Fishes, 317 B.C. to 1844 A.D.							
30 Pisces	45°	42 32	11 17 W	11 W	11 49 W	1 07.5 E	0 00
0 Pisces	75	73 43	42 28 W	39 W	40 22 W	28 52.5 W	27 54W
Aquarian Age began 1844 A.D. (Davidson)							
30 Aquarius	75°	73 43	42 28 W	39 W	40 22 W	28 52.5 W	27 54W
15 Aquarius	90	90 00	58 45 W	-	-	-	-
0 Aquarius	105	106 17	75 02 W	69 W	69 24 W	58 52.52 W	57 48 W
30 Capricorn	105	106 17	75 02 W	69 W	69 24 W	58 52.5 W	57 48 W
0 Capricorn	135	137 28	106 13 W	101 W	101 51 W	88 52.5 W	90 00 W
30 Sagittarius	135	137 28	106 13 W	101 W	101 51 W	88 52.5 W	90 00 W
0 Sagittarius	165	166 11	134 56 W	133 W	134 20 W	118 52.5 W	122 12W
30 Scorpio	165	166 11	134 56 W	133 W	134 20 W	118 52.5 W	122 12W
15 Scorpio	180	180 00	148 45 W	-	-	-	-
0 Scorpio	195	193 49	162 34 W	163 W	162 00 W	148 52.5 W	152 06 W
30 Libra	195	193 49	162 34 W	163 W	162 00 W	148 52.5 W	152 06 W
0 Libra	225	222 32	168 43 E	169 E	168 20 E	178 52.5 W	180 00
30 Virgo	225	222 32	168 43 E	169 E	168 20 E	178 52.5 W	180 00
0 Virgo	255	253 43	137 32 E	141 E	140 33 E	151 07.5 E	152 06 E
30 Leo	255	253 43	137 32 E	141 E	140 33 E	151 07.5 E	152 06 E
15 Leo	270	270 00	121 15 E	-	-	-	-
0 Leo	285	286 17	104 58	111 E	110 36 E	121 07.5 E	122 12 E
30 Cancer	285	286 17	104 58	111 E	110 36 E	121 07.5 E	122 12 E
0 Cancer	315	317 28	73 47 E	79 E	77 05 E	91 07.5 E	90 00 E
30 Gemini	315	317 28	73 47 E	79 E	77 05 E	91 07.5 E	90 00 E
0 Gemini	345	346 11	45 04 E	47 E	45 40 E	61 07.5 E	57 48 E

Note: Table of Right Ascensions is based on Formula: tan R.A. = Cos O.E. tan Long.

The writer uses the precessional chronology of Davidson in preference to that of the other authorities listed on page 14, because of Davidson's meticulous mathematical derivation of the precessional formula from the geometry of the pyramid, checked against Prof. Simon Newcomb's formula given in the *American Ephemeris* and *Nautical Almanac*. Davidson's starting date of 4600 B.C. also agrees with that given by R. Brown, Jr., in his *Primitive Constellations*. Although Councel and Spencer base their chronology on Councel's assumption that the pyramid was oriented to Aldebaran at the Vernal Equinox of 3240 B.C. in 15° Taurus, no mathematical proof is given. The writer however, bases his 15 Taurus orientation of the Pyramid which is located 31°08′ east of Greenwich on the following:

(1) The pyramid is located at the exact center of the land area of the world; therefore, it should, by analogy, be oriented to the middle of the Taurean Age–15° Taurus.

(2) The pyramid is the repository of the highest mental accomplishments of the Egyptian civilization, which is the middle of the three geographical areas covered by the Taurean Age.

(3) The middle decan of Taurus is the Virgo decan during which Egypt's greatest contributions to civilization were made.

The Age of Taurus the Builder, ca. 4699 B.C. to 2508 B.C.

The foundations of our Western Civilization were laid in the East, in the great river valleys of Mesopotamia and Egypt. The earliest historical period of which we have any written record dates back to the 5th millennium B.C., when the development of writing enabled kings to have their names recorded on monuments during their lifetimes. Enough of these monuments and tablets have been discovered to enable historians to piece together a fairly accurate picture of the Euphratean, Egyptian, and Aegean civilizations that developed almost contemporaneously between the meridians 45E and 17E during the Taurean Age, and lasted from ca. 4699 to 2508 B.C.

Each precessional age is dominated by the characteristics of the constellation to which it is oriented. Thus, Taurus is a fixed earth sign; hence the predominant occupations of the people were agricultural. Their arts and sciences were of a practical nature. Since Taurus is the sign of the Builder, we find great construction achievements, such as the Ziggurats of Sumeria (forerunners of the skyscrapers of New York), the temples and pyramids of Egypt, the labyrinth and palaces of Crete. Taurus is also a money sign hence we find great accumulations of wealth in the hands of the rulers. The religions of the age featured the Assyrian Bull, Horus the Strong Bull, and Ptah the Apis-Bull in Egypt, the Aegean legends of the Minotaur of Crete, and Europa and the Bull.

If we divide the 30° of Taurus into three decans of 10° each, covering a span of some 730 years, we find that in harmony with the westward movement of the equinoctial point, the Capricorn element predominates in the Euphratean civilization, the Virgo element in the Egyptian, and the Taurus element in the Aegean. Let us know examine each of these civilizations in a little more detail and see how faithfully the equinoctial clock has recorded the westward march of civilization.

I. The Euphratean Civilization

(a) ca. 4699 B.C. to 3969 B.C. Decan of Capricorn the Organizer—Leading Earth Sign. About 5000 B.C., the Sumerians, users of metals and possessors of a high culture, entered the valleys of Mesopotamia, where they conquered the simple peasant folk. The flatness of land, the absence of natural boundaries, the desirability of uniform control over irrigation, the need for places of refuge against periodic floods and protection from enemies without, let the Sumerians to build cities with strong walls and monumental buildings called ziggurats, on mounds raised above the level of the inundations. These factors were also impelling reasons for the establishment of strong, central governments.

(b) ca. 3969 B.C. to 3239 B.C. Decan of Virgo the Reasoner—Mutable Earth Sign. Although agriculture was the chief industry, commerce with distant lands flourished. The Sumerians were instrumental in establishing commercial and banking practices, forms of written contracts, and were the first to codify civil law; prices and wages were sometimes fixed by law. The oldest inscriptions are on stone, dating from about 3600 B.C., but cuneiform writing on clay tablets was developed out of the earlier pictographs about 3200 B.C. Standard weights and measures were introduced, and the sexagesimal system of computation devised. This system uses 60 as a base, and from it developed our 60 seconds in a minute, 60 minutes in an hour, and the 360o circle. The study of the stars later developed into Chaldean astrology, with six signs of 60° each. A calendar of 12 months was established, based on the phases of the Moon, with the addition of a month whenever necessary to bring the year into harmony with the seasons.

(c) ca. 3239 B.C. to 2508 B.C. Decan of Taurus the Builder—Fixed Earth Sign. Sumer and Akkad were united under Sargon I (ca. 2872-2817 B.C.). Great libraries of clay tablets had been formed by 2700 B.C. The jewelry, cups, swords, and other objects found in the tombs of the kings of Ur by Prof. C. Leonard Woolley of the University of Pennsylvania, show that Sumerian art was at its zenith in the third millennium B.C. King Guden (ca. 2600 B.C.) and his successors built lavishly and beautified their cities with temples. But increasing wealth and the enervating effects of the climate on the inhabitants combined to make men content with what they had and did under the protection of the ancient gods, whose priests had acquired extraordinary power and wealth. Factional jealousies among the petty rulers of the cities resulted in short-lived empires, and civil wars so weakened Sumer that by the end of the Taurean Age, it fell prey to the Elamites from the east and rising Amorites from the west.

II. The Egyptian Civilization

(a) ca. 4699 B.C. to 3969 B.C. Decan of Capricorn the Organizer—Leading Earth Sign. A thousand miles west of Sumer, in the Valley of the Nile, the Egyptians were developing a parallel civilization. But, whereas the Sumerians built with Sun-dried mud and clay, the Egyptians built with enduring stone. Egypt had a delightful climate, was relatively isolated by seas and deserts, and thus protected from invasion. Therefore, the peaceful arts and sciences developed earlier and more rapidly than in Sumer. Knowledge was put to practical use, hence the architects applied their knowledge of surveying and mechanics to the measurement of land and crops, and the building of palaces, temples and tombs. The brilliant skies of day and night favored the development of astronomy, which played an important part in religious matters for fixing the dates of festivals and determining the hours of the light.

Through the observation of the heavenly bodies in the skies and the recurrence of the Nile floods, the Egyptians arrived at the concept of the year. Since the agricultural seasons depended more immediately on the Nile than on solar movements, the first day of the month of inundation, nominally the beginning of the rise of the Nile, was the beginning of the year. The Nile began to rise very rapidly at about the date (July 19 Julian) of the heliacal rising of the Dog Star Sothis (our Sirius) at Memphis, so the Egyptian astronomers took that event as the starting point for the seasonal year, which consisted of twelve months of thirty days each, with five feast days at the end. But the 365-day year was one-quarter day short, so it lost one day in four years, a month in 121 years, and a whole year in 1,461 years; hence the priests arranged their sacred calendar so that at the end of each 1,460 years, a Sothic cycle, the year and the heliacal rising of the Dog Star were brought into agreement. However, the calendar in common use was never corrected. On the basis that the heliacal rising of Sirius coincided with the Egyptian calendar in A.D. 139, E. Meyer calculated that the calendar was introduced in Egypt in 4241-4238 B.C.

(b) ca. 3969 B. C. to 3239 B.C. Decan of Virgo the Reasoner—Mutable Earth Sign. According to Prof. James A. Breasted of the University of Chicago, Menes,[1] the founder of the First Dynasty of the Old Kingdom succeeded in uniting the kingdoms of the north and south under his ruler ship in about 3400 B.C., building a new capital at Memphis. The growth of knowledge was stimulated and the increasing prosperity of the unified kingdom permitted the inauguration of large scale construction of temples, palaces and tombs, the greatest of which were the rock hewn tombs of the kings of the 1st and 2nd Dynasties at Abydos. At the beginning of the 1st Dynasty, there was sudden influx of new ideas in sculpture as in other fields.

---

[1]The two leading schools of Egyptologists disagree in their dating of the reigns of the kings of Egypt. Thus, the founding of the 1st Dynasty by Menes is given as 3400 B.C. by Meyer, Sethe, Breasted, Erman and Steindorff. Other dates are: Boehn, 5702; Unger, 5613; Petrie (in 1906), 5510; Mariette, 5004; Petrie (in 1894), 4777; Brugsch, 4400 Lepsius, 3892; Bunsen, 3623. The chronology of the *Cambridge Ancient History* has been used by the writer.

©) ca 3239 B.C. to 2508 B.C. Decan of Taurus the Builder—Fixed Earth Sign. Imhotep, the architect of King Zoser (ca. 3150 B.C.), the first King of the 3rd Dynasty, designed the oldest Egyptian structure extant, the step-pyramid of Sakkarah, and founded a school of architecture, which provided the next dynasty with the first great builders of history. In about 3000 B.C., the Egyptians invented hieroglyphic writing and used papyrus instead of the cumbersome clay tablets of the Sumerians. The highest point of power and prosperity during the Old Kingdom was reached in the period of the kings of the 4th Dynasty (ca. 2900-3750 B.C.). Khufu (Cheops) (ca 2898-2875 B.C.) for whom the Great Pyramid of Giza was said[1] to have been built, Khafru (ca. 2867-2811 B.C.) of the second pyramid, and Menkaure (ca. 2811-2788 B.C.) of the third pyramid. These monarchs were obsessed by such a passion for making monuments for themselves as no men before or since have had a chance to display or gratify. Sculpture and painting had been brought to heights that later generations of Egyptians could scarcely equal.

While pyramids had been built before and others were built after the 4th Dynasty, they were smaller than the three at Giza, which reached the acme of grandeur and were more complete in sculpture and detail. A total of 100 royal pyramid tombs of the Old Kingdom have been discovered in six groups. These vast structures were erected in an age when engineering science had scarce begun, and so exhausted the resources of Egypt thru three long reigns that it left her wasted as if by war. Meanwhile, the power of the nobles had been growing and that of the king declining, so that after the death of Pepi II (ca. 2538-2444 B.C.), who reigned 94 years, an ominous gap in the monuments points to civil wars and anarchy as having marked the end of the Old Kingdom.

The Aegean Civilization

One hundred eighty miles north of Egypt, 110 miles from Asia Minor, and 60 miles off the southern coast of Greece, facing almost east and west, commanding the Aegean Sea, lays the island of Crete. Here developed a civilization younger indeed than that of Sumer and Egypt, but destined to become second only to Egypt in wealth. Its artists, who were never equaled until the best days of classical Greece, erected the most luxurious palaces in history. Set in the midst of the sea within easy sail of three continents, Crete was influenced by, and influenced them all in turn.

(a) ca. 4699 B.C. to 3969 B.C. Decan of Capricorn, the Organizer—Leading Earth Sign. The original population of Crete came in mast less boats from Asia Minor. Later, before Menes

---

[1]There is likewise little agreement among authorities as to the date of the building of the Great Pyramid of Giza. Davidson, the greatest living Pyramidologist, states that the alpha Draconis orientation to the entrance passage has been calculated as follows: Prof. Piazzi Smith, Astronomer Royal of Scotland, 2170 B.C.; Sir John Herschel, 2160 B.C.; Davidson, 2144 B.C.; Richard Proctor, 2140 B.C., Dr. Brunnow, Astronomer-Royal of Ireland, 2136 B.C. Dr. Brunnow has also calculated an earlier alpha Draconis alignment at 3443 B.C. Davidson concludes: "Until, however, it can be proved from Egyptological or other archaeological data that the reign of the Great Pyramid King Khufu, included the year 2144 B.C., the matter must remain in abeyance."

united Egypt, large numbers of Egyptians and Libyans migrated to Crete and mingled with the earlier Anatolians, developing around 4000 B.C. an elaborate Bronze Age Civilization. Alt6hough the basis of its economy was agriculture, the Cretans excelled as seamen, and were skilled in crafts, especially pottery.

(b) ca. 3969 B.C. to 3239 B.C. Decan of Virgo the Reasoner—Mutable Earth Sign. From Egypt came the first seeds of civilization, for when Menes in 3400 B.C. conquered the Kingdom of the North, those men of the Delta region who would not submit to the victor sailed away to Crete. Because of its central position, Crete became the chief carrier of the trade of Egypt, Syria, Mesopotamia and Greece. As her wealth increased, she became the cultural leader among the islands and coasts of the Aegean. The key to the Cretan system of writing has only recently been discovered by the Swedish archaeologist Prof. Axel Perssons; hence we do not as yet know the name of a single Cretan king nor a single date in native Cretan history.

©) ca. 3239 B.C. to 2508 B.C. Decan of Taurus the Builder—Fixed Earth Sign. Using the name of the legendary King Minos, Sir Arthur Evans the discoverer of the Cretan civilization calls this period of Cretan history Early Minoan II with a tentative dating of 3000-2400 B.C. Her isolation in the sea saved Crete from invasion and allowed her to develop her new-found culture without outside interference. In consequence, she transmuted all she touched into newer, more perfect forms. Cretan civilization is impressive in its originality, in its perfect adaptation to the needs of its people. Her expanding sea-borne trade increased; from the older civilizations of the east she drew constant inspiration, hence during the period of anarchy following the decline of Sumer and Egypt, Crete was laying the foundations for the spread of a luxurious civilization over the barbaric lands to the west.

Summary

While it is not definitely known which of the two lands, Sumer or Egypt, was first to attain high civilization, since they both appeared on the historic scene simultaneously, Sumer was rich in material wealth and art at least 500 years before the first Pharaoh. Will Durant lists the following "firsts" for the Euphratean civilization: Agriculture and trade, irrigation, organization into states and empires, government and law, law code, business contracts, written records, libraries and schools, use of gold and silver as standards of value, credit system, and use of ornamental metal. These are characteristic of the Capricorn decan.

Egypt on the other hand was first to: develop a firm united government, build in stone, devise an effective calendar, develop the sciences, build great ships to sail the seas, create an art that could represent the human figure with accuracy and beauty, express its spiritual life in a religion that promised immortality of the soul in a land beyond the grave. The Egyptians were the greatest builders and sculptors in history. There are no finer specimens in the history of sculpture than the limestone statue of Zoser (ca. 3150 B.C.), the diorite statue of Khafre (ca. 3067 B.C.) and the alabaster statue of Menkaure (ca. 3011 B.C.) in the Cairo Museum. Egypt is definitely the best representatives of Taurus the Builder, and the greatest element in Egyptian civilization was its art, as reflected in its architecture and sculpture, thus corresponding with the Virgo decan.

Contemporary with the period of greatest glory in both Sumer and Egypt, the Cretans, compelled by their geographical position to become seamen, maintained a constant trade with Egypt and Syria. With startling rapidity, the brilliant islanders assimilated the lessons taught them by Egypt, and in a lesser degree, by Asia Minor, transformed foreign models into new and more beautiful patterns, and built up a civilization of their own, which was transmitted to the later Greeks. Thus, the Taurus decan, which saw Sumer and Egypt reach the heights of wealth and power, only to fall into anarchy, witnessed the heartening spectacle of the torch of civilization being carried aloft by Crete.

Chapter 3

The Age of Aries the Warrior
ca. 2508 B.C. to 317 B.C.

Just as the Age of Taurus was pre-eminently that of the Builder, since during it were laid the foundations of our Western civilization, so may the Age of Aries be called that of the Warrior, for it is replete with wars of conquest and of annihilation. The constellation of Aries is said to be dominated by Mars, the god of war, therefore, we could expect nothing else in this Age but force and violence. Let us now divide the 30° of Aries into its three decans of 10° each, Sagittarius, Leo, and Aries, and see how closely history has followed the westward movement of the equinoctial point during its 720-year stay in each decan. We will begin with the civilizations that were developed in the Taurean Age, since there is a marked overlapping of the effects of the various ages on the civilizations that are in existence at the time, the trend, however, being east to west.

I. The Euphratean Civilization–ca. 2508 B.C. to 317 B.C.
(a) ca. 2508 B.C. to 1778 B.C. Decan of Sagittarius, the Prophet–Mutable Fire Sign. About B.C. 2357, Ur of the Chaldeans was sacked by the Elamites of the east, and for the next 200 years nothing much is heard from this area until Hammurabi (ca. B.C. 2123-2080), King of Babylon, restored order by conquering both Sumer and Elam. He established the Code of Hammurabi in 2090 B.C. from a compilation of earlier Sumerian law codes. (The original cylinder was unearthed at Susa in 1902 A.D.) Hammurabi's Code became the basis of the Mosaic Law, which has become part of our own laws. The Babylonians were the first people to write from left to right. Astronomy was the special science of the Babylonians, for which they were famous throughout the ancient world. The Babylonians studied the stars, not so much to chart the course of caravans and ships, as to divine the future fates of men; they were astrologers first

and astronomers afterward. As far back as B.C. 2000, they had made accurate records of the heliacal rising and setting of the planet Venus, which were recorded on the Venus Tablets of Ammizaduga, King of Babylon. They had fixed the positions of various stars, and were slowly mapping the skies. But after Hammurabi's death, a slow disintegration set in, hastened by a Kassite raid in B.C. 2072, and a devastating raid from the Hittites in B.C. 1926. Darkness fell upon the scene until the Kassites under Gandash set up a new dynasty in B.C. 1746.

(b) ca. B.C. 1778 to B.C. 1040 Decan of Leo the Adventurer—Fixed Fire Sign. While the Kassite Dynasty, established in B.C. 1746, nominally ruled over Babylon for 570 years, it gradually receded from the world stage. Its sole contribution to western civilization was the horse, which speeded up communication over land. Assyria, some 300 miles to the north began to gain strength during its long struggle with the Hittites (ca. B.C. 1900-1200) in Anatolia and Mitanni (ca. B.C. 1475-1275) in Upper Mesopotamia. Mitanni was destroyed by Shalmaneser I in B.C. 1275, leaving no perceptible mark on civilization. The Hittites, after having been a thorn in the sides of Egypt and Assyria for some 700 years, were destroyed in B.C. 1200 by the Achaeans. The Assyrians having built up a powerful military machine turned south and broke the power of the Kassites in Babylon in 1174 B.C. Tiglath-Pileser I (ca. B.C. 1116-1093) gained control of the main trade routes of western Asia. The Assyrians, however, suffered a relapse following a surprisingly successful raid from Babylon in B.C. 1107, and sank into temporary oblivion.

c) ca. B.C. 1040 to B.C. 317 Decan of Aries the Warrior—Leading Fire Sign. Not until B.C. 911 were the Assyrians under Adadnirari II able to build a new empire, based on a policy of deliberately destroying the independence of the conquered people. They created the most perfect military machine yet known, and increased its mobility by the introduction of cavalry. They developed an unprecedented grasp of military strategy and tactics, and their invention of the battering ram enabled them to break down the strongest walls. Ashur-nasir-pal II (ca. B.C. 883-859) ruthlessly extended Assyria's power to the Mediterranean. Later rulers extended her power eastward to Damascus, but between B.C. 782 and 745, Assyria lost most of its conquests through incompetent rulers and the first phase of the Assyrian Empire came to an end.

Tiglath-Pileser III (ca. B.C. 745-727) started the second phase of the Assyrian Empire by winning back the lost conquests and extending his power to the borders of Egypt. He consolidated his gains by deporting entire populations. Sagon II (ca. B.C. 722 -705) captured Samaria in B.C. 722 and took the "10 Lost Tribes of Israel" into captivity. Sennacherib (ca. B.C. 705-681) defeated Judah but failed to take Jerusalem. In B.C. 689, he destroyed Babylon with the greatest savagery. Essarhaddon (ca. B.C. 681-668) conquered Egypt in B.C. 671. Ashurbanipal (ca. B.C. 668-625) sacked Thebes in B.C. 661, and destroyed Elam in B.C. 640. He assembled a great library of cuneiform tablets at Nineveh, many of which are now in the British Museum. Despite the fact that the government of Ashurbanipal was without doubt the most extensive administrative organization yet seen in the Mediterranean or near eastern world, fourteen years after his death, Nineveh was destroyed (B.C. 612) by Cyaxares, King of Media and Nabopolassar, King of Babylonia, founder of the Neo-Babylonian (Chaldean) Empire. In B.C. 605, Assyria the Terrible, the eternal type of bestial force, of ruthlessness systematized, was utterly blotted out. Nebuchadnezzar (ca. B.C. 605-561), King of Babylonia, defeated Necho of Egypt at Carchemish in B.C. 605, and captured Jerusalem twice (B.C. 597 and 586), taking the Jews into captivity. Babylon fell to Cyrus, King of Persia in B.C. 538, and together with the kingdom of the Medes and the Persians to Alexander the Great in B.C. 331.

We have thus seen the rise and fall of all the nations that formed the Euphratean civilization; Sumer and Akkad, Mitanni and the Hittites, Elam, Assyria, Babylonia, Chaldea, the Medes and the Persians. Let us now see what was happening in Egypt during this age.

II. The Egyptian Civilization—ca. B.C. 2508 to B.C. 317

(a) ca. B.C. 2508 to B.C. 1778 Decan of Sagittarius the Prophet—Mutable Fire Sign. We have previously seen that as the glory of the Old Kingdom faded with the death of Pepi II of the 6th Dynasty, there ensued a period of disintegration and chaos when weak Pharaohs were unable to maintain a strong, central government. Not until B.C. 2375, with the founding of the 11th Dynasty, was order restored under a central government. With the 12th Dynasty which, was inaugurated by Amenemhet I (ca. B.C. 2212-2192), Egypt entered the second great period of its history—the Middle Kingdom. During it, all the arts, excepting perhaps architecture, reached a height of excellence never equaled in known Egypt before or since. But thirteen years after the death of Amenemhet III (ca. B.C. 2061-2013), Egypt was plunged into disorder, and the Middle Kingdom ended in two centuries of turmoil and disruption (B.C. 2000-1800).

(b) ca. B.C. 1778 to B.C. 1040 Decan of Leo the Adventurer—Fixed Fire Sign. The Hyksos, nomads from Asia, invaded disunited Egypt, set fire to the cities, razed the temples, squandered the accumulated wealth, destroyed much of the accumulated art, and held the Nile Valley in subjection for two hundred years (B.C. 1800-1600). They introduced the horse and chariot into Egypt, where, later, they were used to good advantage.

Ahmose (ca. B.C. 1580-1557) drove out the Hyksos and established the 18th Dynasty, which was to lift Egypt to greater wealth, power and glory than ever before. Now began the thousand year struggle between Egypt and Asia, and during it, the 1st Empire extended from the Nile to beyond the Euphrates. Its greatest leader was Thotmes III, the "Napoleon of Egypt" (ca. B.C. 1479-1447), the first man in known history to recognize the importance of sea power. During the reign of Menhotep III "The Magnificent" (ca. B.C. 1412-1376), the empire attained its greatest splendor and 18th Dynasty art reached its zenith. But the religious revolution and impractability of Akhnaton (ca. B.C. 1375-1358) permitted the Asiatic conquests of his predecessors to slip away from Egyptian control. Akhenaton introduced monotheism into Egypt, but upon his death, the powerful priesthood restored the worship of the ancient gods. The 18th Dynasty ended with the 1st Empire no longer an empire.

A general named Harmhab (ca. B.C. 1346-1322) ushered in the 19th Dynasty by restoring internal order, and thus paved the way for the 2nd Empire. Seti I (ca. B.C. 1321-1300) reestablished Egyptian authority in Palestine, Libya, and other portions of the defunct First Empire. Ramses II (ca. B.C. 1300-1233) attempted grandiose conquests in Asia with little outward success, but the magnitude of his building enterprises left nothing to be desired. His treaty with the Hittites (1280 B.C.) is the first formal diplomatic document extant. Mernephthah (ca. B.C. 1233-1223) crushed revolts in Palestine and Libya, but upon his death a period of anarchy ensued. Ramses III (ca. B.C. 1204-1172) of the 20th Dynasty, smashed the raiders called the "Peoples of the Sea," which included Achaeans and Philistines, in one of the earliest recorded naval engagements in history (B.C. 1190). This was Egypt's last hour of glory, for, while she was free from foreign attack, she was being bled white by the exactions of the priests of Amon. By B.C. 1090, Egyptian authority in Asia became purely nominal, and the Second Empire ended in civil war between the priests and the nobles who ruled as the 21st Dynasty.

©) Ca. B.C. 1040 to B.C. 317 Decan of Aries the Warrior—Leading Fire Sign. Evil days were falling on Egypt, which had now passed from a vigorous and splendid meddle age into premature senility. In B.C. 1954, the Libyans invaded Egypt from the West and Sheshonk I, a Libyan soldier, in B.C. 1945 established the 22nd Dynasty which lasted for some 200 years with nothing to show for it. Anarchy and civil war prevailed during the 23rd Dynasty (B.C. 745-718) and the 24th Dynasty (B.C. 718-712). The 25th Dynasty was Ethiopian and lasted from B.C. 7122 to 663, but it fell when Essarhaddon of Assyria made a province of Egypt in B.C. 671 and Ashurbanipal sacked Thebes, the capital, in 661 B.C. Thus the accumulated wealth of a thousand years became the prey of the most merciless people that ever afflicted the East. Psamtik I (ca. B.C. 663-609) rebelled against Assyria and inaugurated the Saitic Revival in art, sculpture, painting, architecture, literature and religion. Necho (ca. B.C. 609-593) attempted to regain Egypt's lost Asiatic empire; he conquered Josiah of Judea in B.C. 609 and marched on to the Euphrates, but was driven out of Asia by Nebuchadnezzar in B.C. 605. His successors were no more successful in their military campaigns, until, finally, Psamtik III was defeated in B.C. 525 by Cambyses of Persia, who put an end to Egyptian independence. In B.C. 332, Alexander the Great made Egypt a province of Macedonia, but upon Alexander's death in B.C. 323, it fell to one of his generals, Ptolemy I. Thus perished the independent life of Egypt, but its civilization has become an imperishable part of the cultural heritage of mankind.

One of the difficulties facing the researcher is the inability of historians to agree on dates, as is evidenced by the following tabulation. An example is the date of the Biblical exodus from Egypt. Davidson gives B.C. 1486, under Pharaoh Mernepthah, but Finegan gives B.C. 1290, also under Mernepthah, while admitting the plausibility of B.C. 1441, during the reign of

Thotmes III. Sir Charles Marston favors 1440 B.C. under Amenhetep II. Engberg suggests that there were two exoduses, one under Joshua, who entered Palestine from the north around B.C. 1400, and the other under Moses, who entered Palestine from the south around B.C. 1200.

Comparison of Chronologies of Egyptian Dynasties

Twelfth Dynasty, Middle Kingdom

	Hammerton Barnes	Langer	Durant	Petrie	Finegan	Davidson
Amenemhet I	2000-1970	2000-1970	2212-2102	-	1989	2036-2006
Senusret I	1970-1935	1980-1935	2192-2157	-	-	2016-1971
Amenemhet II	1935-1903	1938-1903	-	-	-	1974-1936
Senusret II	1903-1887	1906-1887	-	-	-	1942-1923
Senusret III	1887-1849	1887-1849	2099-2061	-	-	1923-1885
Amenemhet III	1848-1801	1849-1801	2061-2013	-	-	1905-1859
Amenemhet IV	1801-1792	-	-	-	-	1863-1854
Sebekneferu-ra	1792-1788	-	-	-	-	1858-1854

Thirteenth, Fourteenth, Fifteenth, Sixteenth Dynasties

Anarchy-Hyksos Invasion	1788-1635	1788	1800-1600	-	1776	-

Seventeenth Dynasty at Thebes, Contemporary with Hyksos

	Hammerton Barnes	Langer	Durant	Petrie	Finegan	Davidson
Sekemenra I	1635-1615	-	-	-	-	-
Sekemenra II	1615-1605	-	-	-	-	-
Sekemenra III	1605-1591	1600	-	-	-	-
War of Independence						
Uazkheperra	1591-1581	-	-	-	-	-
Senekhtenra	1581-1580	-	-	-	-	-

Eighteenth Dynasty, First Empire, New Kingdom

Ahmose	1580-1557	1580-1557	1580	1573-1560	1570	1829-1804
Amenhotep	1557-1541	1557-1536	-	1560-1539	1546	1804-1784
Thotmes I	1541-1501	1536-1520	1545-1514	1539-1514	1525	1784-1768

Thotmes II		1514-1501	1514-1501	-	-	1768-1745
Hatshepsut)	1501-1479	1520-1480	1501-1479	-	-	-
Thotmes III)		1501	-	-	1501-1447	-
Thotmes III	1479-1447	1479-1447	1479-1447	-	-	-
Amenhotep III	1411-1375	1411-1375	1412-1376	1413-1377	1413	1656-1625
Akhnaton						
Heretic King	1375-1358	1375-1358	1380-1362	1377-1361	-	1625-1607
Smenkhkara	1358	-	-	-	-	1607-1595
Tutankhamen	1358-1353	-	1360-1350	-	-	1595-1586
Ay	1353-1350	-	-	-	-	1586-1573

Nineteenth Dynasty

Harmhab	1350-1321	1350-1315	1346-1322	-	-	1573-1569
Ramses I	1321-1320	-	-	1318-1317	-	1569
Seti I	1320-1300	1313-1292	1321-1300	1317-1295	1319	1567-1558
Ramses II	1300-1225	1292-1225	1300-1233	1295-1229	1301	1558-1491
Mernepthah	1225-1215	1225-1215	1233-1223	1229-1210	1234	1491-1483
Amenmeses	1215	-	-	-	-	1483
Saptah	1215-1209	-	-	-	-	1483-1477
Seti II	1209-1205	-	1214-1210	1210-1205	-	-

20th Dynasty

Setnekht	1200-1198	-	-	-	-	-
Ramses III	1198-1167	1198-1167	1204-1172	-	-	-
Ramses IV to XI	1167-1094	1167-1090	-	-	-	-

III. The Aegean Civilization, ca. B.C. 2508 to B.C. 317

(a) ca. B.C. 2508 to B.C. 1778 Decan of Sagittarius the Prophet—Mutable Fire Sign. While the two earlier civilizations were having their ups and downs, the Cretans were steadily expanding their civilization and establishing colonies on the mainland of Greece and Asia Minor. The historical periods are known as Early Minoan III (ca. B.C. 2400-2200), Middle Minoan I (ca. B.C. 2200-2100), and Middle Minoan II (ca. B.C. 2100-1800). Between B.C. 2200 and B.C. 1800 came the great Age of Invention, which saw the introduction of the four-wheeled cart from Babylonia, the building of stone palaces—notably that of Knossos (ca. B.C. 2150), the development of a cursive style of writing, fresco painting, long bronze swords, etc. Cretan sculpture rose far in advance of any of its contemporaries; its artisans were superior to most of those of Egypt and the Euphratean nations. However, severe earthquakes destroyed Knossos in B.C. 1800 and temporarily arrested further progress.

(b) ca. B.C. 1778 to B.C. 1040 Decan of Leo the Adventurer—Fixed Fire Sign. A speedy recovery from the disastrous effects of the earthquake of B.C. 1800 took place, and the age that followed saw the climax of Cretan glory. But a second earthquake in B.C. 1580 shook the spirit of the people so that when the mainland colonies revolted in B.C. 1400 and invaded Crete, Knossos was destroyed and its treasures pillaged. For a brief while, Cretan civilization was succeeded by that of Mycenae on the mainland, but it lacked the vigor of the Mother Island.

In B.C. 1200, the Achaeans (the first of the true Greeks) swept down from the north and destroyed Mycenae and the Hittites, but they and the rest of the "Sea Raiders" were defeated in B.C. 1190 by Ramses III, as we have previously seen. One of the groups of Sea Raiders—the Philistines, fled into Palestine, where they became a thorn in the sides of the Israelites. Others of the Sea Raiders settled in Italy. In B.C. 1180 the Achaeans crossed the Aegean and destroyed Troy. The Achaeans were suppressed about B.C. 1100 by the Dorian Greeks, another group of the Indo-European invaders that had been moving westward and into northern Greece since about B.C. 2000. Chaos followed the Durians invasion, lasting till about B.C. 1000. The blending of Indo-Europeans with the Mediterraneans is thus believed to have produced the Greek people.

(c) ca. 1040 B.C. to 317 B.C. Decan of Aries the Warrior—Leading Fire Sign. As a result of the Dorian Invasion in B.C. 1100, there arose a three-fold division of the Greek people: (1) the Durians in the south, of which Sparta was the greatest, (2) the Ionians in the middle, of which Athens was the greatest, and (3) the Aeolians in the north. Between B.C. 900 and 600, monarchies were replaced throughout Greece by aristocracies, with the nobles becoming the dominant power in the city-states. Greek and Phoenician colonies were established throughout the Mediterranean (the Greeks in southern Italy and Sicily in B.C. 760), and a thriving sea-borne trade sprang up. Coined money was introduced into Greece from Lydian in Asia Minor about 680 B.C. On May 28, 585 B.C. Alyattes, third king of Lydia and Cyaxares, King of Media, signed a peace treaty as a result of an eclipse of the Sun predicted by Thales of Miletus.

Cyrus the Great, having become ruler of the Medes and the Persians in B.C. 550 defeated Croesus of Lydia in B.C. 546, and Lydia and Ionia became Persian provinces. In B.C. 499 the Ionians revolted against Persia, which at that time, under Darius, covered a wider area than any empire the world had known up to that time. The Persian army was defeated by the Greeks led by Mittiades of Athens at the Battle of Marathon in B.C. 490. Xerxes the Great was defeated by the Athenian fleet under Themistocles at the Battle of Salamis in B.C. 480. The war dragged on until the combined armies of Athens and Sparta decisively defeated the Persians at the Battle of Platae in B.C. 479, while the Athenian navy overwhelmed the remnants of the Persian fleet north of Miletus. This ended the Persian menace, and Athens became the leader of the Greek city-states, reaching its zenith under Pericles (B.C. 460-430). Then began a series of wars—the Peloponnesian Wars—between Sparta, which was a military despotism and Athens, which was democracy with varying fortunes for the principal contenders. Not until the accession of Alexander to the throne of Macedon in B.C. 335 did a Greek leader arise who succeeded in unifying the warring city-states into a single empire. By the time he died in B.C. 323, Alexander the Great had conquered Syria, Egypt, Babylon, Persia, and had penetrated to the Indus in India. But his empire fell apart upon his death, and his generals finally dividing it into three parts—the East, Egypt and Greece.

To the west, in Italy, a new power was rising with the founding of Rome in B.C. 753 as a monarchy under Etruscan kings. About B.C. 500 the monarchy was overthrown and Rome became a republic and the leader of the Latin League. The Sabines and Etruscans in the north were conquered, and then the Greek colonies in the south. By B.C. 275, Rome had become master of all Italy.

On the northern coast of Africa, a Phoenician colony had been founded at Carthage in B.C. 814. Under able leaders, the Carthaginians waged wars of conquest in Africa, Sicily and Sardinia. Developing sea power, it gained control of the western Mediterranean, including the Straits of Gibraltar (ca. B.C. 600-405). The Carthaginians fought four wars against Dionysius I of Syracuse (B.C. 405-367) with varying results. The wars in Sicily were continued intermittently between B.C. 367-268. At one time (B.C. 310-307), Carthage was besieged by the Syracusans.

Summary
In the decan of Sagittarius the Prophet, the Euphratean civilization presented to mankind the religion of the priest kings typified by the Israelites with the symbol of the ram's horn, and the Code of Hammurabi, which was later incorporated into the Mosaic Code and thus into our own laws. Egypt, under the Middle Kingdom, reached its zenith in the arts, while Crete saw the rise of seaborne trade and invention, and reached its zenith in sculpture. These are all characteristics of the sixth natural house ruled by Sagittarius, but since it is a mutable sign, the civilizations of that period did not endure.

In the decan of Leo the Adventurer, we see the rise of new nations dominated by the spirit of adventure and conquest. In the Euphratean Civilization, Mitanni, the Hittites and the Kassites were succeeded by the ruthless power of Assyria. Egypt, under Thotmes III, created the First World Empire reaching the peak of its glory under Amenhotep III, only to sink into decay under the Ramessids. In the Aegean, Crete was destroyed and its civilization passed on to Mycenae, which later was destroyed by the Achaeans, who in turn were conquered by the Dorian Greeks.

In the decan of Aries the Warrior, the true Mars decan, the ruthlessness and senselessness of War reach their peak. Assyria the Terrible, having, by the most barbarous savagery, extended its power over all the Near East, was utterly destroyed by the Chaldeans and the Medes. The Medes were absorbed by the Persians, who then destroyed the Neo-Babylonian Empire and reduced

Egypt to a province. But the vast Persian Empire fell to Alexander the Great, only to break in pieces after his death in B.C. 323, and the Age of Aries ended in chaos. Fortunately for mankind, by the end of the age, the culture of the Orient had been absorbed by the Greeks, whose culture was in turn fused with that of the Orient. The West had finally conquered the East by force of arms, but the cultural heritage of the East lived on in the glory that was Greece.

Chapter 4

The Age of Pisces the Mystic
ca. B.C. 317 to A.D. 1844

We have seen mankind dominated for 2,100 years by the creative urge of Taurus the Builder, and for another 2,100 years by the destructive power of Aries the Warrior, as the equinoctial point moved westward around the ecliptic. The Piscean Age, however, introduces an element of dualism, for the constellation lies over the waters of the Atlantic Ocean. Therefore, the countries bordering on the eastern shore of the Atlantic have been subjected to the joint influence of Aries and Pisces. Hence, war has dominated the peoples of Europe for another 2100 years, but with it have grown two militant religions often engaged in deadly combat—Christianity and Mohammedanism. While both religions originated in the Near East (the home of the earlier Euphratean Civilization), the former made its greatest headway in the West, while the latter held sway over the Middle and Far East. We will now review the history of the nations as the equinoctial point moved through the Scorpio, Cancer, and Pisces decans.

I. Ca. 317 B.C. to A.D. 403 Decan of Scorpio the Alchemist—Fixed Water Sign
(a) The Hellenistic Monarchies. Alexander's sudden death in 323 B.C. left his vast empire without a master. His son was killed in 310 B.C., and the empire was finally divided in 280 B.C. amongst his warring generals who founded the dynasties: Antigonous in Macedonia and Greece; Ptolemy in Egypt; and Seleucus in Asia. As the result of four wars with Rome, Macedonia became a Roman province in B.C. 148, and of Greece in B.C. 146. The Seleucid Empire became embroiled with Rome in 192 B.C. Jerusalem, which had been rebuilt under the rule of the Persians, was destroyed by the Seleucid emperor Antiochus IV in 168 B.C. Media and Persia were lost to the Parthians, and the rest of the once vast Seleucid Empire became a Roman province in 64 B.C. The Ptolemaic Empire, after establishing a brilliant civilization at Alexan-

dria, passed into the hands of Rome in B.C. 31, the dynasty becoming extinct upon the death of Cleopatra in B.C. 30.

(b) The Rise of Rome. In the Second Samnite War (B.C. 326-304) Rome was initially on the losing end, but finally emerged victorious. The Third Samnite War (B.C. 298-290) ended with Rome again the victor, and by B.C. 275 she had conquered all of Italy. But her expansion overseas was challenged by the growing power of Carthage, on the north African coast. The inevitable clash came during the First Punic War (B.C. 264-241) ending in the defeat of the Carthaginians who surrendered Sicily to Rome. The Second Punic War (B.C. 218-201) began disastrously for Rome, but finally ended in the defeat of Carthage which surrendered Spain and the Mediterranean Islands to Rome. Philip V of Macedonia was defeated in B.C. 197, and Antiochus III of Syria in B.C. 190. In B.C. 148 Macedonia became a Roman province, followed by Greece after the ruthless destruction of Corinth in146 B.C. The Third Punic War (149-146 B.C.) ended with the complete destruction of Carthage. In 64 B.C. Pompey made Syria a Roman province; in 55 B.C. Caesar invaded Britain, and in 30 B.C., Octavian made Egypt a Roman province. Rome was now mistress of the world.

In B.C. 4 Jesus of Nazareth was born in the town of Bethlehem, Palestine, which had become a Roman province in B.C. 64. (Ed. Note: There is strong astrological reason to believe that his birth occurred in B.C. 7) In 33 A.D., Pontius Pilate, the Roman procurator of Judea crucified Jesus; in 70 A.D. Titus destroyed Jerusalem, and in 135 A.D. Hadrian dispersed the Jews throughout the Roman Empire. The Golden Age of the Roman Empire occurred between 98 A.D. and 180 A.D., but the 3rd century (192-284 A.D.) was characterized by a complete collapse of government and economics throughout the Mediterranean.

Constantine (324-337 A.D.) reunited the Empire under his sole rule, and in 325 A.D called the first Council of the Christian Church at Micaea, which practically gave official recognition to Christianity as the state religion of Rome. In 330 A.D, he made Constantinople the capital of the Roman Empire, but after his death in 337 A.D., the Empire split in two: the East and West Empires. In the east, the Goths annihilated the army of the Emperor Valens in 378 A.D., and periodically thereafter both the eastern and western Empires were afflicted by Vandal, Visigoth, Ostrogoth, Frank and Hun. Alaric, the Visigoth, who had invaded the Balkans in A.D. 396, was defeated by the Vandal leader of the troops of Honorius, the Roman Emperor of the west, in 403 A.D.

II. Ca. 403 to 1123 A.D, Decan of Cancer the Sustainer—Leading Water Sign
(a) The Western Empire. In 407 A.D. the Romans evacuated Britain and in 410, Alaric sacked Rome. In 435, the Vandal kingdom of Africa was established, and in 450, Attila the Hun invaded Italy, but was defeated at the battle of Chalons in 451. In 455, the Vandals under Genseric sacked Rome, but were finally obliterated by Belisarius in 533. In 476, the Roman Empire of the

west came to an end, with Odoacer, a Teuton, ruling as an independent king as viceroy of the eastern emperor Zeno. In 491, Odoacer was defeated by Theodoric, who set up an Ostrogothic kingdom in Italy which lasted until 552, when it was destroyed by the eastern emperor Justinian (527-565). Italy was ruled from 568 to 774 by the Lombards, who were thereafter incorporated into the Frankish Empire of Charlemagne, which embraced Gaul, Germany, the Netherlands, and Northern Italy.

In 800, Charlemagne was crowned first ruler of the Holy Roman Empire, but following his death in 814, his empire was divided among his three sons, who became the respective kings of the West Franks, East Franks, and of Italy, the latter two eventually being combined. The kingdom of the West Franks ended with the election of Hugh Capet as king of France in 987, but for some 200 years, the French king's authority only extended over a small area around Paris and Orleans. Louis VI (1108-1137) finally became strong enough to defeat an alliance of Henry I of England and the emperor, Henry V, and stopped a German invasion in 1124. The kingdom of the East Franks ended in 911 and was succeeded by Saxon emperors who ruled from 919 to 1024, the greatest of whom was King Otto I (the Great), 936-973, who in 962, revived the Roman Empire in the west. The Salian emperors ruled from 1024 to 1125. The Concordat of Worms, in 1122, ended the troublesome dispute between pope and emperor over lay investiture, which had existed since Charlemagne had been crowned emperor by Pope Leo III in 800. This act affirmed the spiritual supremacy and leadership of the papacy.

(b) Britain. Following the evacuation of Britain by the Romans in 407, the country was subjected to frequent raids by the Picts and Scots on the north, the Irish Celts on the west coast, and Saxon rovers on the east and south coasts.

Christianity was introduced into England by St. Augustine in 597. By 615, the Angles and Jutes had reached the Irish Channel and had become masters of what is now modern England. In 787, the Danes raided England and in 795, invaded Ireland. Alfred the Great defeated the Danes in 878 but had to divide the land with them, he taking the south and the Danes taking the north. His grandson, Ethelstan 924-939), completed the re-conquest of the north from the Danes, but under Ethelred the Unready, England was annexed to the Danish Crown in 1013. William the Conqueror defeated King Harold at the Battle of Hastings, October 14, 1066, and was crowned king of England on Christmas Day 1066. The Norman Conquest was consolidated by William's third son, Henry I (1100-1135), who also settled the question of lay investiture in England.

(c) The Eastern Empire. Attila the Hun, attacked the Eastern Empire in 441, overrunning a great part of the Balkans, but after his death in 453, the leaderless Huns were defeated in A.D. 454 and their power broken. Justinian (527-565) codified Roman law, which became the basis of almost all the legal systems of Europe. His able General, Belisarius (505-565) defeated the Persians in 530, and annihilated the Vandals in 533. Another general, Narses, wiped out the

Ostrogoths in 552. After Justinian's death, the Empire fell on evil days, being invaded by the Persians who sacked Jerusalem in 614, but were finally defeated in 627 at Nineveh by the emperor Heraclius.

From 634 to 641, Syria, Mesopotamia, and Egypt were lost to the Moslems. After a five-year siege (673-678), Constantinople secured a favorable 30 year peace from the Arabs. Between 675 and 681 the Slavs attacked Thrace, Macedonia, and northern Greece. In 697-698, the Arabs brought an end to Byzantine rule in North Africa. Constantinople withstood another siege by the Arabs in 717-718, but in 746 they expelled the Arabs from Anatolia, only to lose it again in 837-838. The Eastern Empire made peace with Charlemagne in 803, thus retaining southern Italy, Venice, and Dalmatia. In 817 the Bulgars were defeated and obliged to accept a 30-year peace.

Basil I (867-886) founder of the Macedonian Dynasty initiated what was probably the most glorious period of Byzantine history. The Byzantine general John Kurkuas extended the empire's power to the Euphrates and Tigris Rivers in brilliant campaigns (920-942). The piratical fleets of Tripoli were defeated in 924, as well as a great armada of Russians in 941. Syria was incorporated in the empire in 99; the Bulgarian army was annihilated in 1014, and the combined Lombards and Normans defeated in A.D. 1018. The final schism between Rome and Constantinople occurred in 1054, followed by the end of Byzantine rule in Italy in 1068. During the First Crusade (1096-1099) the entire western coast of Anatolia was reconquered, Jerusalem was recaptured July 15, 1099, and the Latin kingdom of Jerusalem was founded. By 1109, Caesarea, Tripoli, Tyre and Sidon had been recaptured. The Seljuk Turks were defeated (1110-1117), the Byzantines gaining most of Anatolia and in 1121, southern Anatolia was recovered.

(d) The Rise of Islam. Mohammed (570-632) proclaimed himself the Prophet of God in 612, founding a militant religion that brought all of Arabia under his sway by the time of his death in 632. His successors swarmed out of Arabia, sweeping away age-old empires. Syria fell in 636, Jerusalem in 638, Mesopotamia and Persia by 641, Egypt in 642, northern Africa in 670, northern India to the Jaxartes River in 676. Constantinople was besieged for five years, but a 30-year truce was concluded in 678. The Visigothic Kingdom of Spain was invaded in 711 and with the exception of the mountain region of Asturias, was subjugated by 715. Constantinople was again besieged in 717-718, and southern France was invaded, but the Moslems were finally defeated by Charles Martel at the Battle of Tours in 732, and were pushed back over the Pyrenees by 759. Although the Moslems continued to raid the southern part of Gaul up to the time of Charlemagne, their power in the west slowly receded.

Dynastic dissensions contributed to the decline of the Moslem power--the Omayyad Dynasty ruling from Damascus (661-750) was succeeded by the Abbasid Caliphate, ruling from Bagdad (750-1100), but one of the Omayyads, Abdurrahman (756-787), set up an independent kingdom in Spain, ruling as their Emir of Cordova. Although Charlemagne re-conquered northeastern Spain by 801, Cordova became the greatest intellectual center of Europe during the reign of Abdurrahman III (912-961). Much of what is now Portugal was re-conquered from the Moslems by Ferdinand the Great of Leon and Castile in 1055. Toledo was recaptured by Alfonso VI of Castile in 1085.

The Abbasid Caliphate again attacked Constantinople, forcing the Empress Irene to sue for peace (783-785). In the reign of Mamoun the Great (813-833), the most glorious epoch in the history of the Caliphate, two observatories were built, one near Bagdad and the other near Damascus; literary, scientific, and philosophical works were translated from Greek, Syrian, Persian, and Sanskrit. The Moslem Empire lost its predominantly Arab character, and commenced that remarkable assimilation of Persian, Byzantine, and Hellenistic culture for which it became so noted during the Middle Ages, and which was to prove so important in the cultural progress of medieval Europe.

Crete was conquered in 825, Sicily in 827, Palermo in 831. War with the Byzantines was renewed (837-842), and Rome was sacked in 846. The Abbasid Caliphate was overthrown by the Seljuk Turks, who had been converted to Islam in the 9th and 10th centuries, and who captured Bagdad in 1055. The Seljuk Turks then conquered Georgia and Armenia by 1073 and virtually destroyed the power of the Byzantines in Asia Minor in 1071. But in 1092, the Seljuk Empire was rent by civil wars among the captains of the western coast of Anatolia (1096-1099) and were defeated by the Byzantines (1110-1117), abandoning most of Anatolia, and losing southwestern Anatolia in 1121.

II. ca. A.D. 1123 to A.D. 1844 Decan of Pisces the Mystic—Mutable Water Sign

(a) The Eastern Empire (Byzantine Empire). War with Venice (1122-1126) resulted in a Venetian victory. War with Roger of Sicily ended in an inconclusive peace in 1158. The Second Crusade (1147-1149), which achieved nothing of importance and which was discredited throughout Europe, almost brought on war with Constantinople. In 1176, Constantinople paid a heavy indemnity to Venice, and the entire Byzantine empire began to fall to pieces (1185-1195). The Fourth Crusade (1202-1204) was diverted by the Venetians against Constantinople, which was stormed and sacked with unparalleled horrors by the crusaders on April 12, 1204. They were there upon excommunicated by the pope.

The Byzantine Empire was then divided equally between the Venetians and the crusaders, Count Baldwin of Flanders becoming the first of the Latin emperors. The Latin Empire ended with the recapture of Constantinople by the Greeks in 121. The empire of the Paleologi (1261-1453), while undergoing a territorial and political decline, was accompanied by an extraordinary cultural revival analogous to the Renaissance in Italy. The Civil War (1341-1347) proved the undoing of the Empire, since both sides called in Serbs or Turks to support them. The

Ottoman Turks thus were able to establish themselves in Europe at Gallipoli in 1354. Constantinople finally fell to Mohammed, the Conqueror, on May 29, 1453, thereby bringing an end to the 1,100-year-old Eastern Empire.

(b) The Moslems. The Seljuk sultanate at Bagdad ended in a disastrous collision with the Eastern Turks in 1159. The western sultanate at Damascus became embroiled with the Caliphate of Egypt, becoming victorious in 1169 A.D. Their great leader, Saladin (1138-1193), captured Jerusalem in October 1187 and destroyed the Latin Kingdom which had been established there in 1099. The Third Crusade was thereby precipitated, but rivalries among the Christian rulers wrecked it, and King Richard (the Lion Hearted) of England was forced to accept a three-year truce with Saladin under which Christian pilgrims could enter Jerusalem freely. The Fifth Crusade (1218-1221) resulted in a 10 year truce with Emperor Frederick II of Germany being crowned King of Jerusalem in 1229. In 1243 the Mongols captured Anatolia from the Seljuks, and in 1258, they sacked Bagdad, but were defeated by the Mamelukes in Egypt in 1260. Jerusalem was sacked by Mongol mercenaries in 1244, but remained in Moslem hands until its recapture by General Allenby in 1917.

In Spain, the Moslem power was slowly waning. They were expelled from Castile in 1212 A.D. by Alfonso VIII of Castile, from the Balearic Isles (1232) and Valencia (1238) by James I of Aragon. Ferdinand III of Castile recaptured Cordova (1236), Seville (1244), Jaen (1246), Cadiz and Xeres (1250). Alfonso XI of Castile decisively defeated the Moslems in 1340. Grenada fell to Ferdinand and Isabella in November 1491, and the Moors were expelled from Granada in 1502, thus ending their almost 800-year occupation of Spain.

The Ottoman Turks (1300-1481), who had made their first European settlement at Gallipoli in 1354, made Adrianople their capital in 1365. They conquered Bulgaria (136-1372), Serbia (1371-1386) and invaded Greece (1397). Constantinople was besieged (1391-1398) and forced to pay heavy tribute. The Mongols under Tamerlane invaded Anatolia in 1400 and shattered the Ottoman Army in 1402. During the first war with Venice in 1416, the Turkish fleet was destroyed off Gallipoli. In the next war (1425-1430), the Turks were victorious. John Hunyadi defeated the Turks in Hungary in 1442 and freed Wallachia and Serbia in 1443. Constantinople was captured May 29, 1453 (the day having been set by Mohammed's astrologers), and became the capital of the Ottoman Empire. Although the Turkish advance on the Danube was halted by Hunyadi's heroic defense of Belgrade in 1456, Serbia was re-conquered (1456-1458) and Bosnia and Herzegovina (1458-1461).

The Wars of Venice (1463-1479* and 1499-1503) ended in favor of the Turks. Selim I (1465-1520 defeated Persia (1514), took Syria and the Mamelukes (1516) and annexed Egypt (1517). Suleiman the Magnificent (1520-1566) captured Belgrade (1521), Rhodes (1522), but was repulsed from Vienna in 1529. The Holy League against the Turks ended in the defeat of the

Venetians in 1540. The War with Persia (1553) ended in 1555 with Suleiman retaining his Mesopotamian conquests. The Turks took Cyprus (1570) but suffered a crushing defeat at the hand of Don John of Austria at the naval Battle of Lepanto, October 7, 1571. In the War with Poland (1672-1676) the Turks gained the Ukraine, but lost it to the Russians in their First War with Russia (1677-1681).

In the war with Austria (1682-1699) the Turks lost Hungary, Transylvania, Croatia, and Slavonia to Austria; Morse and Dalmatia to Venice; Podolia to Poland, and Azov to Russia. The Siege of Vienna was relieved by Jan Sobieski in 1683. The Turks regained Azov from the Russians (1714-1718). In the war with Austria and Russia (1736-1739) the victorious Turks regained Serbia, but allowed Russia to retain Azov. In the First War with Catherine the Great of Russia (1768-1774), Russia conquered the Crimea, but the Turks regained Moldavia and Wallachia. The Second War with Catherine the Great (1787-1792) was inconclusive. Napoleon failed to conquer Egypt from the Turks (1798-1799). The Serbs revolted (1804) and gained their freedom (1815-1817). In the War against England and Russia (1806-1812) Bessarabia was given up to Russia. The Greek War of Independence (1821-1830) freed the Greeks. The conflict between the Sultan of Turkey and Mohammed Ali of Egypt (1832-1833) ended in Mohammed gaining Syria. A gentlemen's agreement between Russia and England regarding the precarious condition of the Ottoman Empire was concluded in 1844.

(c) France (Kingdom of the West Franks). The first monarch to increase the prestige of the crown was Louis VI (1108-1137), who added Aquitaine through the marriage of his son to the heiress of Aquitaine. Philip II (1180-1223) acquired Normandy, Maine, Anjou, and Touraine from the English (1202-1204). A brilliant cultural advance accompanied the general material and political progress of the time of Philip II and Louis IX (1226-1270). Philip VI (1328-1350) was the first ruler of the Capetian House of Valois and during his reign the Hundred Years' War (1338-1453) with England began. The English defeated the French at the Battles of Crecy (1346), Poitiers (1356) and Agincourt (1415), but were defeated by Joan of Arc at the Battle of Orleans (1429), and finally were expelled from most of France (1449-1461). Calais, the last English possession in France, was captured by the Duke of Guise in 1558. The Religious Wars (1562-1628) were highlighted by the Massacre of the Protestants on St. Bartholomew's Day, August 23-24, 1572, the Edict of Nantes (1598), which granted equal political rights to the Huguenots, and the Siege of La Rochelle (1627-1628), which ended in the subjugation of the Huguenots.

Cardinal Richelieu (1624-1642) unified France, restored French influence in Italy, Germany, and the Netherlands, established it in Sweden, and laid the foundation for the power of Louis XIV (1643-1715). Louis' ambition for fame and his desire for increase in territory involved France in war with: Spain over the Spanish Netherlands (1667-1668); Holland (1672-1678); the League of Augsburg (1688-1697) brought on by the Revocation of the Edict of Nantes in 1685;

Spain-War of the Spanish Succession (1701-1714). The Battle of Blenheim (1704), however, put a stop to Louis XIV's territorial expansion. Law's Mississippi Scheme (1718-1720) brought widespread financial disaster to France. The War of the Austrian Succession (1740-1748) involved France in conflict with England in North America and India. The Seven Years' War (1756-1763) led to the loss of most of the French Colonial Empire. Louis XVI (1774-1792) helped the American colonists in the War of American Independence (1776-1783 but as a result of the French Revolution (1789), he lost this throne September 21, 1792, and his head on January 21, 1793. The Reign of Terror (1793-1795) ended with the beginning of the rise to power of Napoleon Bonaparte.

Napoleon's Italian campaign (1796-1797) was successful, but his Egyptian expedition (1798-1799) was a failure, for his fleet was destroyed by Nelson at the Battle of the Nile, August 1, 1798, and his army had to be repatriated after being defeated by the English at Alexandria, March 21, 1801, practically involved the destruction of the 1000-year-old Holy Roman Empire. The Treaty of Amiens, March 27, 1802, between England and France achieved a complete pacification of Europe. Napoleon became emperor of the French (1804-1814) and was soon at war with the rest of Europe. At the Battle of Trafalgar, October 21, 1805, the naval power of France was broken. But on land Napoleon was successful in battle after battle until the disastrous retreat from Moscow, October 19, 1815. The tide had turned and on April 11, 1814, Napoleon abdicated unconditionally and was exiled to the Island of Elba. During the deliberations of the Congress of Vienna, September 1814-June 1815, Napoleon escaped from Elba and began the Hundred Days Rule, which ended with his defeat at Waterloo, June 18, 1815. This time he was banished to the Island of St. Helena where he died in captivity in 1821, and France was deprived of all the conquests he had made for her in twenty years of warfare. Louis XVIII (1814-1824) became king of France upon the abdication of Napoleon, and succeeded in paying off the indemnity levied on France by the allies. After the Paris Revolution of 1830, Louis Philippe (1830-1848) ascended the throne, but failed to institute needed reforms. The growth of industrialism stimulated socialism and racialism, and led to the Revolution of 1848.

(d) Germany (Kingdom of the East Franks) Holy Roman Empire. Lothair II (1125-1137) was the first German king to ask papal approval of his election. Upon his death, the House of Hohenstaufen (1138-1268) became the first German dynasty committed to a policy of aggrandizement of the lay imperial power, which precipitated a second great struggle between the popes and the emperors. Frederick Barbarossa (1152-1190) regarded himself as heir to the tradition of Constantine, Justinian and Charlemagne, and aimed at restoring the glories of the Roman Empire. Frederick II (1211-1250) negotiated a 10-year truce with the sultan of Egypt and had himself crowned king of Jerusalem (1229). He was in frequent conflict with Popes Gregory IX and Innocent IV, being twice excommunicated. The Great Interregnum (1254-1273) marked the end of the medieval Holy Roman Empire and the failure of imperial efforts to establish German unity. Rudolf I (1273-1291), the first of the Hapsburgs, founded the greatness of the dynasty on territorial expansion of the family holdings and dynastic marriages. He threw away the last remnants of Frederick II's great imperial fabric: Confirmation of papal rights in Italy and Angevin rights in Southern Italy (1275); renunciation of all imperial claims to the papal States and Sicily (1279). The Diet of Frankfurt (1338) divorced the Holy Roman Empire from the papacy. The Golden Bull (1356) transformed the empire from a monarchy into an aristocratic federation, to avoid the evils of disputed elections. Frederick II (1440-1493) was the last emperor crowned at Rome by the pope.

The Diet of Worms (1495) attempted the modernization of the medieval empire. The Diet of Cologne (1512) carried the reorganization of the empire further. Martin Luther (1483-1546) began the Reformation of October 31, 1517. Charles I of Spain was chosen emperor by the German electors and became Charles V (1519-1556). He defeated Francis I of France in two wars (1521-1526 and 1527-1529) but was stalemated in the next two (1536-1538 and 1542-1544). He was the last German emperor crowned by the pope, being crowned at Bologna in 1530. Upon his abdication in 1556, the crown of Spain with Naples, Milan, Franche-Comte, and the Netherlands went to his son Philip, the imperial office and the Hapsburg lands to his brother, Ferdinand I. The Thirty Years' War (1618-1648) developed into a general conflict between Catholic and protestant Europe, and political struggles by Sweden and France on German soil resulted in irreparable losses of men and wealth.

Leopold I (1658-1705) engaged in war against the Turks (1661-1664), liberated Hungary from the Turks (1682-1699), and joined in the war of the League of Augsburg (1688-1697) against Louis XIV of France. As a result of the war of the Spanish Succession (1701-1714) Austrian power and influence replaced that of Spain in Italy and the Netherlands. In the War of the Polish Succession (1733-1735) and in the war with the Turks (1736-1739), the Austrians were unsuccessful. The War of the Austrian Succession (1740-1748) brought Prussia, under Frederick the Great (1740-1786), into the ranks of the Great Powers. The Seven Years' War (1756-1763) resulted in the defeat of the Austrians at the hands of Frederick. In the First Partition of Poland (1772), Austria gained Red Russia, Galicia, Western Podolia, Lemberg, and part of Cracow; Prussia gained Polish Prussia. In the War of the Bavarian Succession (1778-1779), Austria was again the loser.

The alliance of Austria and Prussia against France (1792) led to the Napoleonic wars which comprised: the War of the First Coalition (1792-1797), the War of the Second Coalition (1798-1801), and the War of the Third Coalition (1805-1812), during which the old Holy Roman Empire ended on August 6, 1806. The Austrians and Prussians were usually on the losing end until the final defeat of Napoleon at Waterloo (1815). In the Second Partition of Poland (1793) Prussian gained Danzig, Thorn and Great Poland. In the Third Partition of Poland (1795) Prussian gained Nazovia and Warsaw, while Austria obtained the remainder of the Cracow region.

The end of the Napoleonic wars resulted in a reactionary Germanic Confederation brought into being by Prince Metternich of Austria at the Congress of Vienna (1815). Frederick William IV of Prussia (1840-1861) dreamt of reviving the Holy Roman Empire with Prussia playing a glorious role, but secondary to that of the Hapsburgs. In 1842 he called an inconclusive meeting of representatives of the Provincial Diets to discuss popular representation in the government of Prussia. The radicals in Hungary, led by Louis Kossuth, sought independence from Austria and parliamentary government for Hungary. In Austria, the masses were goaded into riots by a severe economic depression which led to the outbreaks of 1848.

(e) Great Britain. The death of Henry I in 1135 ushered in a dynastic war and a period of feudal anarchy. Pope Adrian IV gave Ireland to Henry II in 1154. In wars with Philip II of France, England lost most of her French possessions (1202-1204). The barons forced King John to sign the Magna Charta at Runnymede, June 15, 1215 (For chart of this event see *Astrological Americana, Journal of Research #3*, American Federation of Astrologers). Edward I (1222-1307) conquered Wales (1276-1284) and Scotland (1285-1307), removing the Coronation Stone from Scotland to England (1296). At the Battle of Bannockburn (1314), Robert Bruce defeated the English and became king of Scotland, thereby postponing for centuries the ultimate union with England. The Hundred Years' War (1338-1453) over the French possessions of the English crown resulted in initial English victories at Crecy (1346), Poitiers (1356), and Agincourt (1415), during which period most of France fell into the hands of the English. The tide turned with the defeat of the English by Joan of Arc at Orleans (1428-1429), and by 1453 only Calais remained in English hands.

The English were constantly embroiled with the Irish (1315-1485). The Wars of the Roses (1455-1485) resulted in the victory of the House of York over the House of Lancaster, but upon the death of Richard III at Bosworth Field, the crown passed to Henry VII of the House of Tudor. Henry VIII (1509-1547) broke with the Church of Rome (1529), and parliament by the Act of Supremacy (1534) gave official recognition to the English Reformation. Ireland was made a kingdom (1542) and Calais was captured by the French (1558).

During the reign of Queen Elizabeth (1558-1603), Catholics were persecuted, the Anglican Church was established (1563), Sir Francis Drake circumnavigated the globe (1572-1580), Sir Walter Raleigh founded the first Virginia Colony (1585), the Spanish Armada was destroyed (1588). An Irish Rebellion was put down (1597), and the East India Company was chartered (1600). During the reign of James VI of Scotland, who became James I of England (1603-1625), the first permanent English Settlement in America was made at Jamestown, Virginia (1607) (For chart of this event see *Astrological Americana, Journal of Research #3*, American Federation of Astrologers): Henry Hudson discovered the Hudson River (1609), the Pilgrims landed at Plymouth (1620) (ibid.), and the Dutch founded New Amsterdam (1623). During the reign of Charles I (1625-1649) who was embroiled with parliament and beheaded on January 30, 1649, the Great Migration to Massachusetts Colony (1630-1642) took place. The Commonwealth (1649-1660) was established, with Oliver Cromwell as Lord Protector.

During the reign of Charles II (1660-1685), New Amsterdam was taken from the Dutch, August 27, 1664, and renamed New York; the Great Plague (April 1665) and the Great Fire (September 2-9, 1666) decimated most of London; the Hudson Bay Company was incorporated (1670); colonies were established in the Carolinas (1670) and Pennsylvania (1681); and the Habeas Corpus Act was passed (1679). In the reign of William and Mary (1689-1702) American colonists were massacred at Schenectady (1690) by the French and Indians during the King William's War (1689-1697)—the American phase of the War of the League of Augsburg against Louis XIV; the Bank of England was chartered (July 27, 1694), and the London Stock Exchange was formed (1698). During the War of the Spanish Succession (1701-1714) the English captured Gibraltar (July 24, 1704), and the Duke of Marlboro defeated the French and Bavarians at the Battle of Blenheim (August 13, 1704), thereby breaking the power of Louis XIV. In the American phase of this war, known as Queen Anne's War (1702-1713), Acadia was captured from the French and was renamed Nova Scotia.

England and Scotland were united under the name of Great Britain May 1, 1707. The bursting of the South Sea Bubble (1720) brought on a disastrous financial panic. Georgia was founded (1733) and England was engaged in three wars with Spain: The Triple Alliance (1717-1718); the War of Jenkins Ear (1739-1741) in which Georgia was attacked by the Spanish and Florida by the English. In the War of the Austrian Succession (1740-1748), in the American phase of which, known as King George's War (1743-1748), Louisburg was captured from the French (1745). During the reign of George III (1760-1820) the American Colonies gained their freedom in the War of American Independence (1776-1783). The French Republic declared war on Great Britain (1793) and the United Kingdom of Great Britain and Ireland was established on January 1, 1801. The United States declared an inconclusive war on Great Britain (June 18, 1812), during which Washington was burnt by the British in 1814. In the Napoleonic Wars, Nelson defeated the French fleets at the Battle of the Nile (1798) and the Battle of Trafalgar (1805), and Wellington ended Napoleon's career at the Battle of Waterloo (June 18, 1815).

A long and severe economic depression brought widespread distress in Britain after the Napoleonic Wars ended. The Catholic Emancipation Bill (1829) removed most of the political disabilities of the Catholics. Parliamentary reforms were obtained through three reform bills (1830-1834) and trade unionism took a spurt (1834). During the reign of Queen Victoria (1837-1901), a rebellion broke out in Canada (1837) which was ended by the Union Act of 1840. By the Ashburton Treaty of 1842, Britain abandoned to the U.S. more than half the territory claimed on the northeast frontier. The Bank Charter Act of 1844 restricted the issuance of bank notes to the Bank of England. A gentlemen's agreement (1844) was entered into with Russia regarding Turkey—the "sick man of Europe."

(f) Russia. Russia, in the 12th century, was a loose federation of states, with Kiev as its political center. The first mention of Moscow occurred in 1147. The Teutonic Knights (founded 1190-1191 in Palestine) were commissioned to conquer and convert Prussia in 1226. The Mongols (1237-1240) swept over southern and central Russia and into Europe, capturing Kiev (1240) and establishing themselves on the lower Volga (1242). For two centuries thereafter, the Khanate of the Golden Horde ruled all Russia. Alexander Nevski (1236-1263), Prince of Novgorod, defeated the Swedes (1240) and the Teutonic knights (1242). Ivan I (1325-1341), Grand Prince of Moscow, was placed above the other Russian princes by the Tatar overlord. Ivan the Great (1462-1505), regarded as the first national sovereign of Russia, annexed most of the rival principalities, married Sophia, niece of the last Greek emperor of Constantinople (1472), threw off the Tatar yoke (1480) and gained many of the border territories of White Russia and Little Russia from Lithuania (1503).

Ivan the Terrible (1533-1584) conquered Kazan and Astrakhan from the Tatars (1552-1556), granted trade rights to the English (1553), conquered part of Livonia (1563), ravaged Novgorod (1570), but saw Moscow sacked by the Crimean Tatars (1571), was defeated by the Swedes (1578) and the Poles (1581). The Time of Troubles (1604-1613) was filled with conflicts amongst the nobles, Cossacks, Poles and Swedes. Order was restored by Michael Romanov (1613-1645) the founder of the Romanov Dynasty, which ruled Russia till 1917. During his reign, Russian pioneers reached the Pacific (1637) and the Cossacks took Azov from the Crimean Tatars (1637). In the war with Poland (1654-1667), Russia gained the Smolensk region and the eastern Ukraine. A great peasant revolt led by the Don Cossacks was suppressed with great difficulty (1670-1671). In 1681 the Turks abandoned most of the Turkish Ukraine in Russia. Peter the Great (1689-1725) gave up the Amur region of China (1689), captured Azov (1696), was defeated by Charles XII of Sweden (1700), and moved his capital to St. Petersburg (1703). At the Battle of Poltava (July 8, 1709) Peter broke the power of Charles, and Russia succeeded Sweden as the dominant power in the north. In the war with Turkey (1710-1711), Azov was returned to the Turks. By the Treaty of Nystedt (1721) between Russia and Sweden, Russia acquired Livonia, Estonia, Ingermanland, part of Karelia, and a number of Baltic islands.

In the War of the Polish Succession (1733-1735) the Russians prepared the way for the Partition of Poland. In War against the Turks (1736-1739) the Russians recaptured Azov (1737). The Russians gained part of Finland in the war with Sweden (1741-1743). In the Seven Years' War (1756-1763) Russia fought on the side of Austria and France against Frederick the Great of Prussia, but made no direct gains. Under Catherine the Great (1762-1796) theRussians gained unprecedented victories in war with the Turks (1768-1772). In the First Partition of Poland (1772) Russia gained White Russia and all the territory to the Dvina and Dnieper Rivers. In 1783 Russia annexed the Crimea and in the second war of Catherine against the Turks (1787-1792), she gained the boundary of the Dniestr River. In the Second Partition of Poland (1793) Russia took Lithuania and most of western Ukraine, including Podolia. In the Third Partition of Poland (1795), Russia took what remained of Lithuania, Courland, and the Ukraine, and thus became an ever more important factor in European affairs. Russia participated in the war of the Second Coalition against France (1799-1801), her General Suvarov several times defeating the French in Italy.

In War with Persia (1804-1813), Russia gained Georgia, Daghestan, and Shemakha. From 1805-1812 Russia expanded in Alaska and Northern California. In the war of the Third Coalition against France (1805-1807) Russia was defeated and Alexander I (1801-1825) became the ally of Napoleon. In the war against Turkey (1808-1809) Russia acquired Finland. Napoleon's invasion of Russia (1812) was a failure and upon the downfall of Napoleon, Russia gained the Grand Duchy of Warsaw at the Treaty of Vienna (1815). Russia gained Armenia and Erivan in War with Persia (1826-1828) and the mouth of the Danube and the eastern coast of the Black Sea in war against Turkey (1828-1829). The Polish Revolution (1830-1831) collapsed. A Russian squadron arrived in Constantinople (1833) and Russia signed the Straits Convention of 1841. Czar Nicholas visited London and made a gentlemen's agreement (1844) with Lord Aberdeen about what should be done in case of a Turkish collapse.

(g) The New World Discoveries. The practically simultaneous discoveries of the ocean routes from Europe to the east and to the New World in the west resulted inevitable from the "Great Navigations" inaugurated by Prince Henry the Navigator (1394-1460) of Portugal, who built (1418) an observatory on the headland of Sagres, one of the promontories that terminate at Cape St. Vincent, the extreme southwest point of Europe. Here he set up a school of seamanship, employing Arab cartographers and Jewish astronomers from the Moorish universities of Cordova, Seville, and Toledo to prepare maps and nautical tables, to make instruments, and to instruct his captains and assist in piloting his vessels during the next 40 years. The Madeira Islands were explored (1418-1419), the Azores discovered (1427-1431), Cape Verde rounded (1445), the Cape Verde Islands discovered (1455-1457), the Gold Coast discovered (1470-1471), the Congo River and Cape St. Augustine (1482-1484), Bartholomew Diaz rounded the Cape of Good Hope (1487-1488), and Vasco da Gama discovered the direct route to India via the Cape of Good Hope (1497-1499).

Although there are evidences that Norsemen under Erik the Red discovered Greenland (985-986), Leif Ericsson discovered Vinland (Nova Scotia, 1000), Thorfinn Karlsefni spent three winters (1003-1006) on the American continent, Pining and Pothorst reached Greenland (1470-1474), the real discovery of America is credited to Christopher Columbus, who landed on the Bahamas October 12, 1492 and then discovered Cuba and Santo Domingo. On his second voyage (September 25, 1493-June 11, 1496), Columbus discovered Dominca, Puerto Rico, and other islands of the Antilles group. On the third voyage (May 30, 1498-November 25, 1500), Columbus discovered Trinidad Island and the mouth of the Orinoco River in South America. On his fourth voyage (May 11, 1502-November 7, 1504), Columbus reached the coast of Honduras and passed south to Panama. John Cabot sailed to Newfoundland (1497) and New England (1498). The Portuguese Gaspar de Corte Real voyaged to Greenland and Labrador (1500); Cabral discovered Brazil (1500) and Amerigo Vespucci sailed south along the Brazilian coast (1501-1502). Ponce de Leon discovered Florida (1512), Balboa discovered the Pacific (1513), Juan Diaz de Solis explored the coast of South America from Rio de Janeiro to the Rio de la Plata (1515-1516), Cordoba discovered Yucatan (1517), Alvarez Pineda explored the Gulf of Mexico from Florida to Vera Cruz (1519), Ferdinand Magellan circumnavigated the globe (1519-1522), Esteban Gomez explored the American coast from Nova Scotia to Florida (1524-1525).

(h) Spanish America. Under the Bull of Partition of Pope Alexander VI, May 4, 1493, all lands to the south and west toward India, beyond a line drawn one hundred leagues west of the Azores and the Cape Verde Islands, not held by a Christian prince on Christmas Day 1492, were granted to the Catholic kings. By the Treaty of Tordesillas, June 7, 1494, between Portugal and Spain, the line of demarcation was moved 270 leagues further west—Portugal to have all lands to the east of it, and Spain all lands to the west. Brazil was thus brought into the Portuguese sphere. By 1600 the territory from New Mexico and Florida on the north to Chile and the Rio de la Plata on the south, with the exception of Brazil, was under the rule of Spain. Friction between Portugal and Spain was not eliminated by the Line of Demarcation, open warfare resulted in the War of the Seven Reductions, 1752-1756).

Following the discoveries of Columbus, Santo Domingo became the first seat of Spanish government in the West Indies, negro slavery being introduced in 1501. Puerto Rico was conquered (1508-1511), Cuba conquered (1509-1513), Panama was colonized (1509-1523), Mexico was conquered (1518-1522), Guatemala and Salvador conquered (1523-1525), Nicaragua and Honduras conquered (1524-1526), Columbia and Venezuela colonized (1525-1538), Yucatan conquered (1527-1535). On the west coast of South America, Pizarro conquered Peru (1531-1533), Chile was conquered (1535-1553). On the east coast of South America, Sebastian Cabot explored the Parana and Paraguay Rivers (1526-1532) and Mendoza founded Buenos Aires (1536) and Asuncion (1537). The first printing press was introduced into Mexico (1535). Moving north from Mexico, Coronado traversed New Mexico, Texas, Oklahoma, and eastern Kansas (1540-1542). Cabrillo explored the Pacific Coast as far as Oregon (1542-1543). The Spanish Inquisition was introduced into the New World (1569), and seriously hampered intellectual activity. Juan de Onate conquered New Mexico and explored the region from Kansas to the Gulf of California (1598-1608). Texas was not permanently occupied till 1720-1722, and upper California until 1769-1786. The mouth of the Columbia River was discovered (1774-1776) and temporary settlements made on Vancouver Island and Cape Flattery (1789-1795).

On the east coast of North America, attempts to colonize Florida (1521-1528) and the Carolinas (1526-1528 and 1559-1561) were unsuccessful. De Soto explored the southeastern portion of the United States, discovering the Mississippi River and traversing Arkansas and Oklahoma to the Arkansas River (1539-1543). St. Augustine, Florida was founded in 1565. During the latter half of the 16th century and most of the 17th century, French, British and Dutch freebooters raided the Spanish Colonies and took heavy toll of Spanish commerce. In the 18th century and early part of the 19th century, British control of the seas and the naval weakness of Spain rendered protection of her colonies increasingly difficult. At the end of the Seven Years' War (1763) Spain ceded Florida to Great Britain and France ceded Louisiana to Spain. At the end of the American War of Independence (1783) Spain regained Florida. During Napoleon's Consulte, France forced Spain to return Louisiana (1800) and Napoleon sold it to the United States (1803). Spain sold Florida to the United States (1819-1821).

Political, economic, and social factors inherent in the Spanish colonial system were fundamental causes that led to the separation of the American Colonies from the mother country. The first serious insurrectionary movements occurred in Paraguay (1721-1725), Peru (1780-1782), Columbia (1781), Venezuela (1790-1806), Mexico (1810), Guatemala, San Salvador, Honduras, Nicaragua, Costa Rica (1811-1814), all of which were suppressed. But the next wave of uprising, during the Napoleonic Wars, was more successful. The freedom of the United Provinces of the Rio de la Plata region was inaugurated by the British, who temporarily occupied Buenos Aires (1806), although independence was not proclaimed until July 9, 1816.

Under the leadership of Jose de San Martin (1778-1850) independence was achieved by Paraguay, August 14, 1811 (which later separated from Argentina in 1813); Chile, February 12, 1818; Peru, July 22, 1821. Uruguay was incorporated into Brazil (1821) becoming an independent buffer state August 27, 1828. Simon Bolivar (1783-1830) was instrumental in freeing Great Colombia December 17, 1819; Bolivia August 6, 1825. Mexico proclaimed its independence February 24, 1821; and the United Provinces of Central America, July 1, 1823.

(i) Portuguese America. The Portuguese, after the discovery of Brazil (1500), began colonization (1530-1532), the first Governor-General making Bahia, which he founded (1549), the seat of government. A French Colony, established (1555) on the Bay of Rio de Janeiro, was destroyed by the Portuguese, who founded the city of Rio de Janeiro (1565-1567). The Dutch captured Bahia (1624) but were forced to return it (1625). From 1630 to 1654, the Dutch captured a large part of Brazil, but were expelled in the latter year. Spain and Portugal were in armed conflict (1680-1683) over the territory around the future city of Montevideo (founded 1723 by the Spaniards). Rio de Janeiro was sacked by the French (1711) during the War of the Spanish Succession (1701-1714). Wars between the Portuguese and the native Paulistas (1708-1709) and the natives of Pernambuco (1710-1711) ended in favor of the Portuguese. By treaties (1750) and 1777) between Spain and Portugal the latter gained extensive areas in the Amazon and Parana River basins. The Jesuits who had entered Brazil in 1549 were expelled from Portugal and her possessions in 1759.

Prince John, Regent of Portugal, fled before a Napoleonic army to Brazil (1807-1808) and established his government at Rio de Janeiro. Brazil became a member of the United Kingdom of Portugal, Brazil and Algarves, December 16, 1815. Prince John captured French Guiana (1808) but was forced to return it by the Treaty of Paris (1814). He sought to extend his rule over the Rio de la Plata Provinces of the Spanish (1816-1821). Extensive intermixture of negro, indian and white blood occurred in Brazil, negroes having been introduced during the middle of the 16th century. The Inquisition was not introduced into Brazil. Dom Pedro proclaimed Brazilian Independence on September 7, 1822 and was proclaimed emperor October 12, 1822. War between Brazil and Argentina (1825-1828) resulted in the formation of the buffer state of Uruguay. Dom Pedro was forced to abdicate (1831) and during the regency (1831-1840) of his son, anarchy reigned and a separatist movement in Rio Grande do Sul (1835-1845) was quelled.

(j) North America (United States and Canada). The French unsuccessfully attempted to colonize the St. Lawrence River Valley (1500-1600). They were similarly unsuccessful in attempts to colonize Florida (1562-1567). Champlain made the first settlements in Canada (1603-1613) founding Quebec July 3, 1608. French settlements were made in the West Indies (1625-1664). Marquette and Joliet voyaged to the Mississippi (1673), La Salle explored the Mississippi Valley (1679-1683) and Louisiana was founded (1699). Detroit was founded by Cadillac in 1701.

Sir Francis Drake, in circumnavigating the globe (1572-1580) claimed California for England. Martin Frobisher explored Labrador and Baffin Land (1576-1578); Sir Humphrey Gilbert claimed Newfoundland (1583), and Thomas Cavendish duplicated Drake's exploit of circumnavigating the globe (1586-1588). After unsuccessful attempts (1584, 1587) Virginia was colonized at Jamestown (May, 1607), the first negro slaves being introduced in 1619. Two unsuccessful attempts to found colonies in New England were made in 1606, another in 1607, one by Captain John Smith of the Virginia Colony (1615), the Pilgrims finally landing at Plymouth (November 1620). Settlement was very rapid during the Great Migration to Massachusetts (1630-1642). Roger Williams was banished from Salem, Massachusetts, and settled at Providence, Rhode Island (1638). Lord Baltimore founded Maryland (1633).

King William's war with France (1689-1697) was inconclusive, but English colonists were massacred at Schenectady (1690). In Queen Anne's War (1702-1713) Acadia--the home of "Evangeline," was captured from the French (1710) and renamed Nova Scotia. In the wars of Jenkins Ear (1739-1743), unsuccessful attempts were made by the English against Florida, and by the Spaniards, against Georgia. In King George's War (1743-1748) Louisburg was captured from the French. During the French and Indian War (1755-1763), all of Canada passed into the hands of the British. The English Colonies expanded across the Appalachian Mountains to Kentucky (1763-1775). As the result of various oppressive measures enacted by the British Government (1763-1775), the thirteen colonies revolted (1775-1783), the Declaration of Independence being signed July 4, 1776.

The end of the Revolutionary War brought on a deep depression (1784-1785); Washington was inaugurated the first president (1789); Alexander Hamilton reorganized the finances (1790-1791); Eli Whitney invented the cotton gin (1792); Louisiana was purchased from the French (1803); Robert Fulton built the steamboat (1807); the War of 1812 was a fiasco; Florida was purchased from Spain (1819); the Monroe Doctrine was proclaimed (1823); the Erie Canal was completed (182). The Mormon Church was organized (1830), the abolitionist movement was organized (1831); McCormick invented the reaper (1834); Colt, the revolver (1836); Deere, the steel plow (1837); Texas proclaimed its independence from Mexico (1836), and the Panic of 1837 created widespread distress. Calhoun signed the Treaty for the Annexation of Texas on April 12, 1844.

Summary of Piscean Age
In the Scorpio decan (317 B.C. to 403 A.D.), the chaotic influence of Neptune is exemplified in the disintegration and death (VIIIth House) of Alexander's Empire; the cruelty of Scorpio and the ruthlessness of Mars and Pluto are manifested in the rise of Rome which utterly destroyed Carthage, Corinth and Jerusalem, and crucified hundreds of thousands of people. Manilius aptly expresses the characteristics of this period in the following:

"Bright Scorpio, armed with poisonous tail, prepares Men's martial minds for violence and for wars. His venom heats and boils their blood to rage, and rapine spreads o'er the unlucky age."

In the Cancer decan (403 A.D. to 1123 A.D.), the inconstancy of the Moon is exemplified in the senseless dynastic struggles of the fragments of the Roman Empire, which aptly enough met its death in the IVth (Cancer) House. This was also the age of religious fanaticism (Piscean), in which the cruelest kind of warfare (Mars) was waged between Pagan and Christian, Christian and Moslem, Catholic and Protestant. Little wonder that this period is known in history as the "Dark Ages.".

In the true Piscean decan (1123-1844 A.D.), the aesthetic element in Neptune bursts forth in the glorious art of the Renaissance, the spiritual element expresses itself in the Reformation. The symbol of Pisces consists of two fishes—one swimming upstream, representing the strivings for the higher self—the other swimming downstream representing the pull of the baser elements. These latter expressed themselves in the horrors of the Spanish Inquisition and the religious bigotry and intolerance of the age. The scientific element of Neptune expressed itself in the development of the natural sciences, and Neptune's rulership of the seas manifested itself in the great navigations—the Age of Ocean Exploration. This led to the discovery and settlement of the New World, where mankind could lay the foundations for a new, democratic way of life—free from the horrors of the Old World. It is not without significance that the great revolutionary movements, i.e. the American Revolution, the French Revolution, and the South American Wars of Liberation took place within 70 years of the End of the Piscean Age in 1844.

Chapter 5

The Aquarian Age
ca. 1844 A.D.–

(a) Religious and Spiritual Movements. In ancient and medieval times, comets were regarded with superstitious dread--they were thought to be the harbingers of war, pestilence and famine; to foretell the dethronement of kings and the dissolution of empires, and were usually associated with some critical period in history. Thus, a great comet appeared in 11 B.C. (the Star of Bethlehem), in 1066 A.D. at the time of the Norma Conquest of England, and another in 1456 A.D., three years after the capture of Constantinople by the Turks. However, in 1682 A.D. Edmund Halley, a contemporary of Sir Isaac Newton, applied the former's law of universal gravitation to the great comet of that year and worked out its orbit (about 75 years). He concluded that the comets of 1066, 1456, and 1682 were different appearances of the same comet, and predicted its return in 1759, 1835 and 1910. His predictions were fulfilled in each case. Search of ancient and medieval manuscripts of Europe, China and Japan enabled Cowell and Crommelin of the Greenwich Observatory to trace 28 observed returns of Halley's Comet with fair certainty to 87 B.C., and with some degree of probability to 240 B.C.

Nevertheless, when a brilliant comet extending a quarter of the way across the sky appeared in February 1843, the Millerites believed that their leader's (William Miller 1782-1849) predication of the end of the world was about to come true. Miller, in 1831, had calculated from the prophecies in the Books of Daniel and Revelation that the Messiah would appear and the world would end on or about March 20, 1843. This was later extended to July 10, 1943 and then March 22, 1844, but no Messiah appeared, and the Earth continued in its accustomed orbit. In Palestine, Joseph Wolf predicted the advent for 1847. In Lebanon, Lady Heater Stanhope, niece of William Pitt, kept two white Arab horses, one for the Messiah and one for herself! The failure of the world to end in 1844 caused students of Biblical prophecy to revise their ideas—now they speak of the "end of the age." Davidson's calculations by the Precessional Formula of the Great Pyramid, places the end of the Piscean Age at January 25, 1844.

The history of mankind has shown that whenever the spiritual life of men has become degenerate and their morals corrupt, that most wonderful and mysterious of men, the Prophet, makes his appearance. Thus, we have seen arise in the east those great world-teachers: Krishna, Buddha, Zoroaster, Moses, Jesus, Mohammed, who proclaimed anew the devine gospel of righteousness and truth. Therefore, the end of the Piscean Age found the world in the death pangs of an era where the old principles of materialism and self-interest, of sectarian, religious, nationalist prejudices and animosities were perishing. Out of the dying embers of the past, a new age—the Aquarian Age—was being born, bringing with it revolutionary changes of unprecedented magnitude in every department of human life.

The harbinger of this new age was Mirza Ali Muhammad (1819-1850), the Bab ("Door of the Spirit") who on May 23, 1844, announced in Shiraz, Persia, the coming of a new world teacher and founded Bahaism. After the Bab's martyrdom in Tabriz, Persia, the leadership of the movement was assumed by Mirza Hussyn Ali (1817-1892), surnamed Baha u'llah ("Glory of God"). Baha u'llah taught that the Holy Spirit had once more come to revivify humanity in its hour of need, that a new and greater cycle had begun—the Age of Brotherhood, of peace, of knowledge of God. Baha u'llah was succeeded by his oldest son Abdul-Baha (1844-1921), ("Servant of the Glory"), who had voluntarily shared the life imprisonment of his father until 1892, and was himself not released until 1908. The movement is now headed by Abdul's grandson, Shoghi Effendi, with headquarters at Wilmette, Illinois, where a unique temple is under construction and is expected to be dedicated in 1951.

In 1844 Brigham Young (1801-1877) succeeded Joseph Smith as president and prophet of the Mormon Church. He led the great Mormon migration to the "Promised Land" (Utah) where he helped found Salt Lake City in 1847. The Promised Land was prophesied to be the safest place on the continent during the great earthquakes and changes that were believed to come in the future.

In 1845, Andrew Jackson Davis, a born clairvoyant, delivered 157 lectures in New York announcing a new philosophy of the universe and predicting a glorious period before mankind. He predicted the horseless carriage, the Pullman car, the diesel locomotive, the airplane, the modern apartment houses, concrete, a kind of material heaven, and "greater leisure throughout America, and the development of those intuitive faculties in man which are now supposed to be merely possibilities."

Between 1840 and 1857, Phineas Parker Quimby travelled throughout New England lecturing and demonstrating his method of hypnotism and mental healing. Following his death in 1866,

Mary Baker Eddy (1821-1910), who had been healed by Quimby in 1862, announced her version of Quimby's teachings as Christian Science, publishing her book *Science and Health with Key to the Scriptures* in 1875. About that same year, Charles and Myrtle Fillmore founded the New Thought Movement in Kansas City. In 1882, Mr. and Mrs. J.A. Dresser, students of Quimby, began practicing and teaching the Quimby Method of mental healing.

In 1875, Helena P. Blavatsky (1831-1891), together with Col. H.S. Olcott and Wm. Q. Judgo, founded the Theosophical Movement in the United States. Rudolf Steiner, in Germany, called his interpretation of Theosophy "Anthroposophy." Max Heindel, in California, called his Christianized version of Theosophy "Rosicrucianism." In his *The Rosicrucian Cosmo-Conception*, Heindel says, "Buddha, great, grand and sublime, may be the 'Light of Asia' but Christ will yet be acknowledged the 'Light of the World'."

I. Europe

Heinrich Heine (1800-1856) wrote in 1842: "When the terrible wheel of revolution begins to turn we shall see this time the emergence of the most appalling of all the antagonists who have stood forth to do battle against the existing order. Communism is the secret name of this redoubtable adversary who sets against the existing bourgeois world the domination of the proletariat with all its consequences . . ." The Communist Manifesto of Karl Marx (1818-1883) and Friedrich Engels (1820-1895), issued in 1848, said "Let the ruling classes tremble at a Communist revolution. The proletarians have nothing to lose but their chains. They have a world to win. Working men all countries unite." The year 1848, known as the "Year of Revolutions," saw the rise of revolutionary movements in France, Germany, the Austrian Empire, Italy, England and Ireland. Within eighteen months they were all suppressed, except in France.

France. Out of the third French Revolution came the Second Republic (1848-1852) with Louis Napoleon, nephew of Napoleon I, being proclaimed president (1848). The Second Empire (1852-1870) was proclaimed (1852) with Napoleon II as Emperor; France declared war on Russia (1854); the War of France and Piedmont against Austria brought Savoy and Nice to France (1860); the Mexican Expedition (1861-1867) ended in disaster, and the France-Prussian War (1870-1871) ended Napoleon's career. The Third Republic (1870-1914) made peace with Prussia (1871), ceding Alsace-Lorraine to the victor. The France-Russian Alliance was concluded (1891-1894). The Panama Canal Scandal (1892-1893) and the Dreyfus Affair (1894-1906) shook France. Church and state were separated (1901-1905) and epidemics of strikes and labor troubles occurred (1906-1911).

Germany. The Revolutionary Movement of 1848 failed by June 1849; Prussia gained Schleswig-Holstein from Denmark (1863-1864) and won the Seven Weeks' War (1866) against the other German States, and the war with Austria (1867), Prussia won the France-Prussian War (1870-1871) and William I was proclaimed German Emperor (1871). Bismarck separated

Church and State (1871), the Socialist Workingmen's Party was formed (1875), the Triple Alliance (Germany, Austria, Italy) was formed (1882), the first German Navy Law (1898) started a naval race with Great Britain, and the Agadir Affair (1911) almost precipitated a European war. Germany declared war on Russia, August 1, 1914.

Austria-Hungary. The Revolution of 1848 in Austria and Bohemia was suppressed (1848) and with the aid of the Russians, the Hungarian Revolution was suppressed (1849). By the Germanization of the Empire (1849-1860) she sought to destroy the various nationalistic movements. Austria lost in the war with Prussia (1867). Austria was given a mandate over Bosnia and Herzegovina (1908), lost to Serbia in the Balkan Wars (1912-1913), and precipitated World War I (191401918) when the Archduke Ferdinand was assassinated, and Austria declared war on Serbia, July 28, 1914.

Italy. With collapse of the Napoleonic Empire (1815), the states of the Italian peninsula were reconstituted under the domination of Austria. The Neapolitan Revolution (1820), the Piedmontese Revolution (1821) and revolts in papal states (1831), were suppressed. The Italian War of Independence (1848-1849) ended in failure. Cavour of Piedmont (1810-1861), with the aid of Napoleon III, defeated the Austrians (1858-1860), Garibaldi (1807-1882) took Naples (1860), and the Kingdom of Italy was proclaimed March 17, 1861, with Victor Emanuel as its first king. Italy declared war on Austria (1866) and annexed Rome (1870), which had been held by the French since 1849. Italy entered the Triple Alliance (1882), obtained a foothold on the Red Sea (1885), and in Ethiopia (1887-1889), suffered one of the worst colonial disasters in modern Ethiopia (1896), and a severe earthquake and tidal wave in Calabria and Sicily (1906). She won the Tripolitan War (1911-1912) with Turkey, and proclaimed her neutrality in World War I (1914).

Russia. The revolutions of 1848-1849 in the European nations brought forth greater repressive measures in Russia, but opposition to the government increased with Russia's defeat in the Crimean War (1853-1856). Russia gained at the expense of China (1858-1860), the serfs were liberated (1861), the Second Polish Revolution (1863-1864) was suppressed, Alaska was sold to the United States (1867), the Russians conquered the Khanates of Kokand, Bokhara and Khiva (1865-1876). Through the Russo-Turkish War (1875-1878) Russia gained Bessarabia, Kars and Batum. The Transcaspian region was annexed (1881), and continued advances brought Russia to the border of Afghanistan (1884-1887). The France-Russian Alliance was concluded (1891-1894), she lost the Russo-Japanese War (1904-1905), an insurrection of workers in Moscow was suppressed (1905), the Anglo-Russian Entente was concluded (1907), Germany declared war on Russia, August 1, 1914.

Turkey. The Crimean War (1853-1856) ended in the defeat of Russia, insurrections in Syria (1860-1861) and Crete (1866-1868) were suppressed, the Suez Canal was opened (1869), Turkey lost Bosnia and Herzegovina to Austria, Bessarabia, Kars and Batum to Russia, and freed Bulgaria during the Russo-Turkish War (1875-1878), and a second insurrection in Crete was suppressed (1878). Turkey lost Tunis to the French (1871) and Egypt to Great Britain (1882), and put down an insurrection in eastern Roumelia (1885-1888). A third insurrection in Crete was suppressed (1889), the Armenia revolutionary movement was destroyed (1890-1897), the Cretan Insurrection (18976-1897) led to the Turco-Greek War (1897), Bulgaria proclaimed its independence and Bosnia and Herzegovina were annexed by Austria (1908). Abdul Hamid was deposed (1909), an insurrection in Albania was suppressed (1910), Turkey lost Tripoli in the Italo-Turkish War (1911) and Crete, Albania, and the Aegean Islands in the First Balkan War (1912-1913). Russia declared war on Turkey, October 30, 1914.

Great Britain. The Oregon Boundary Treaty with the United States was signed (1845), the Corn Laws repealed (1846), the revolutionary movement in Ireland was suppressed (1848), the Crimean War (1854-1856) with Russia was successful, the Indian mutiny was suppressed (1857), War with China (1857-1858) opened treaty ports, and the British East India Company's political powers were assumed by the Crown (1858). The Irish question (1868-1870) was not solved, the Suez Canal shares of the Khedive of Egypt were purchased (1875), wars were waged against the Afghans (1878-1879) and the Zulus in South Africa (1879), the Irish Question again became troublesome (1881-1882). Britain occupied Egypt (1882), re-conquered the Sudan (1896-1898), won the Boer War (1899-1902), granted Home Rule to Ireland (1914), and declared War on Germany, September 3, 1914.

II. The Western Hemisphere
United States. President John Tyler brought about the annexation of Texas by Joint Resolution of Congress March 1, 1845. War with Mexico followed (1846-1848), resulting in the American acquisition of New Mexico and California. The California Gold Rush began January 24, 1848. The Gadsden Purchase of December 30, 1853, rounded out American possessions in the far west. Commodore Perry opened Japan to American trade (1854) and the Panic of 1857 caused widespread distress. Drake's petroleum well near Titusville, Pennsylvania (1859) marked the beginning of the modern oil industry. The Civil War (1861-1865) resulted in the freeing of the negro slaves of the south, and Alaska was purchased from Russia (1867). The Panic of 1873 was precipitated by the failure of Jay Cooke & Co. Increasing immigration of Chinese in California led to the Chinese Exclusion Act (1882). The American Federation of Labor was organized (1886), the Pan American Union was established (1889), the Spanish-American War (1898) brought freedom to Cuba; and Puerto Rico, Guam, and the Philippines to the United States. The Philippine Insurrection (1899-1902) ended in American victory, the Panama Canal rights were secured (1903), the Panic of 1907 ensued, the Federal Reserve Bank Act was passed (1913). Vera Cruz was seized (1914), and a punitive expedition was sent into Mexico (1916-1917).

Canada. The Oregon Boundary Treaty (1846) between Canada and the United States removed a cause of friction; British Columbia was withdrawn from the jurisdiction of the Hudson's Bay Company (1858); the British North America Act (1867) united Ontario, Quebec, New Brunswick and Nova Scotia into the Dominion of Canada, and the Northwest Territories were purchased from the Hudson's Bay Company (1869). The Red River Rebellion (1869-1870) was put down; Manitoba (1870), British Columbia (1871) and Prince Edward Island (1873) joined the Dominion. The Canadian Pacific Railway was chartered (1881), Northwest Rebellion (1885) was quelled; the Canadian Northern Railway chartered (1898); the provinces of Alberta and Saskatchewan were formed (1905), and the Grand Trunk Railway was completed (1914).

Alaska. Alaska was explored by Vitus Bering, a Russian (1728-1741); Captain James Cook made the first English explorations (1776); the Russian-American Fur Company was organized by the Russian government (1799); foreign ships were prohibited by Russia (1821), and the United States purchased Alaska from Russia (1867). An Homestead Law was passed (1903) and a Boundary dispute between the United States and Canada was settled in favor of the United States (1903). Alaska was given Territorial status (1912).

Mexico. The end of the Mexican War brought Santa Ana to the presidency (1853-1855), followed by the War of the Reform (1857-1860). Napoleon III of France sought to create a new empire in Mexico (1861-1867) but was forced to withdraw by the United States, which invoked the Monroe Doctrine. Porfirio Diaz (1830-1915) ruled Mexico with an iron hand (1876-1911), and while vast material progress was achieved, education was neglected, and the working class was exploited. Diaz was overthrown (1911); Madero (1913); Huerta (1914), and the United States was forced to send a punitive expedition into Mexico (1916-1917).

Argentina. In 1852 President Rosas was thrown out. He had ruled since 1829. Buenos Aires did not join the Federal Union until 1862; war with Paraguay occurred (1865-1870); Buenos Aires was defeated in the Civil War of 1880; boundary disputes with Brazil (1895) and Chile (1899, 1902) were adjusted by arbitration. Secret voting and universal suffrage were instituted (1912).

Brazil. Great material progress took place after 1850. She aided in the overthrow of the Argentine dictator (1851-1852), conducted a major war against Paraguay (1865-1870); freed negro slaves (1888). Emperor Dom Pedro II was deposed (1889) and a Republic was proclaimed. A federal constitution was adopted (1891); a rebellion (1893-1895) was suppressed; boundary controversies with Argentina (1895), France (1900), Bolivia (1903), England (1904) and Holland (1906), were adjusted by arbitration.

Chile. Material progress continued and between 1851 and 1871 a shift toward greater democracy occurred. Chile united with Peru in war against Spain (1865) resulting in the bombardment of Valparais (1866); war with Peru and Bolivia occurred (1876-1870 in which Chile gained rich territory. Civil war ensued (1891) and boundary disputes with Argentina were settled (1902) by arbitration. Internal progress took place (1906-1920).

Paraguay. Paraguay prospered under the presidency of Carlos Lopez (1844-1862), but his son (1862-1870) brought on war with Brazil, Argentina and Uruguay, during which (1865-1870) the Paraguayan nation was almost annihilated. The close of war was followed by a long period of political instability (1872-1912).

Uruguay. Uruguay was plagued with civil war (1843-1851), French and British troops occupying Uruguayan territory (1845-1849). Brazilian forces occupied the Uruguayan frontier towns (1864-1865); Uruguay was allied with Brazil and Argentina in the Paraguayan War (1865-1870), and political instability with many changes in the presidency continued through the first decade of the 20th century.

Bolivia. Between 1847 and 1857 Bolivia was plagued with political instability, lost territory to Chile (1866) and Brazil (1867). In the War of the Pacific (1879-1884) Bolivia lost rich nitrate territory to Chile, and in 1903 ceded additional territory to Brazil. The War of the Pacific was formally terminated by a treaty between Bolivia and Chile on October 20, 1894. Economic development and material progress continued (1909-1913).

Peru. Peru emerged from Civil War (1842-1845) into a period of progress (1845-1862). War with Spain (1864-1865) and (1866) resulted in new presidents. During the War of the Pacific (1879-1884) Chile was victorious and gained rich nitrate territory, and Tacna and Arica were ceded for 10 years. The war brought Peru to the verge of national collapse and a long period of attempted readjustment followed (1885-1912).

Ecuador. Ecuador went through a period of political instability and foreign complications (1845-1860). She united with Peru and Chile against Spain (1865-1866). A period of political instability (1875-1895) was followed by the growth of liberalism to 1924.

Colombia. Colombia granted the United States the right of transit across the Isthmus of Panama (1846); Slavery was abolished (1851-1852; Civil War occurred (1861 and 1867-1880). The French were given a 99-year concession to build a canal across Panama (1878). Liberal revolts occurred (1885 and 1899-1900), but the Civil War of 1900-1903 ended in Victory for the Conservatives. The French Company, which had begun work on the panama Canal (1881), became bankrupt (1891) and its rights were awarded to the United States (1903) for 40,000,000 by Panama, which had revolted. The resulting dispute with the United States was not settled until 1921.

Venezuela. Venezuela abolished slavery (1854); prospered (1870-1888); became embroiled in a boundary dispute with Great Britain (1895-1896); was subjected to a naval blockade by Great

Britain, Germany, and Italy (1902); suspension of diplomatic relations with the United States and a blockade by Holland (1908).

Central America. Attempts to form a confederation of El Salvador, Honduras, Costa Rica, Nicaragua, and Guatemala (1849-1852) (1862) (1876) (1889) (1895-1898) all failed. After a war between Honduras and Nicaragua (1907) a Central American Court of Justice was established (1908) and when Nicaragua ignored the Court's decision in 1917, its influence waned.

(c) Political and Social Movements, 1914-1950. Seventy years after the Aquarian Age began, the world was plunged into the greatest convulsion known in the history of mankind–World War I (1914-1918), during which a vast area of Europe was devastated beyond recognition, 13,000,000 lives were lost and succeeding generations saddled with debts totaling $181 billion for direct costs and $152 billion for indirect costs. This war, between the Central Powers—Germany, Austria-Hungary, Bulgaria, and Turkey—and the Allied Powers—Great Britain, France, Russia, Italy, Japan, and the United States, etc.—started out with initial successes for the Central Powers but ended in their complete defeat. The political systems of Europe were shattered; the Hohenzellern Dynasty in Germany, the Hapsburg in Austria-Hungary, the Romanoff in Russia and the Osmanli in Turkey, were all destroyed; Russia overthrew czarism only to become enslaved by Bolshevism; a host of new states were created and the territorial chaos was incomparably greater than at the end of the Napoleonic wars.

The Armistice of the next 20 years (19191-1939) saw attempts to establish collective security through the League of Nations and the World Court; the efforts of Japan to dominate the Far East by force, and of the United States to maintain her leadership in Europe, and the efforts of the Germans to revise or evade the terms of the peace treaties; the efforts to restore world trade and general prosperity in a war-impoverished world culminating in the Panic of 1929 and the Great Depression of the 1930s; the decline of democracy, self-government an civil liberties, as Bolshevism, Fascism, and Nazism rose to power.

World War II (1939-1945) between the Fascist Powers (Germany, Italy, and Japan) and the United Nations (Great Britain, France, United States, China, Communist Russia, etc.) opened with initial successes for the Fascist or Axis Powers, but ended in their complete defeat. This war took the lives of 22,000,000 people, including civilian, the military cost was $1,116 billion and the property damage was $231 billion. The United Nations was formed on October 24, 1945, with 51 nations as charter members, in the hope of avoiding an even more disastrous global conflict in the future. Unfortunately, the intransigence and growing imperialism of Communist Russia has thus far proved an effective stumbling block, and the Cold War between the Democracies and Communist Russia and the satellites she dominates, threatens at any moment to break out into World War III. Poland, Finland, Hungary, Romania, Yugoslavia, Albania, Czechoslovakia, and China have fallen under the domination of the Soviet. India, Burma, the

Philippines, Palestine, and the Dutch East Indies have gained their freedom, but there is no peace.

(d) Scientific and Technological Developments. The first hundred years of the Aquarian Age have seen a tremendous growth in science and invention of which the following list is but a partial summary:

1844	Samuel F.B. Morse transmits the first telegraph message
	Dr. Horace Wills demonstrates use of nitrous oxide as an anesthetic
1845	Joule proves heat a form of energy
1846	Neptune discovered
	Dr. W.T.G. Norton demonstrates use of ether as anesthetic
	Elias Howe invents the sewing machine
1847	Von Helmholtz announces theory of conservation of energy
1851	Scott Archer invents wet collodion process in photography
	Charles Page builds first electric locomotive
1852	Henry Giffard builds first practical dirigible balloon
1855	Whitworth and Armstrong apply principle of rifling to artillery and small arms
1856	Bessemer develops Bessemer process of steel making
1856-7	Development of dry collodion process in photography
1859	Drake discovers oil at Titusville, Pennsylvania
	Charles Darwin writes *Origin of Species*
1860	Winchester introduces repeating rifle
1862	Gatling invents machine gun
	The first engine to use commercial gas was invented
1864	George Pullman builds first railroad sleeping car
1865-66	First transatlantic cables laid
1865	Lister initiates practice of antiseptic surgery
	Mendel establishes the mechanics of heredity
	William Bullock invents web printing press
	Pierre Martin develops open hearth process of steel making
1866	Robert Whitehead invents locomotive torpedo

1868	Scholes typewriter first used
1869	Westinghouse air-brake patented
	Refrigeration in railway transportation introduced
	First color photograph introduced
	Suez Canal opened
1871	R.L. Maddox invents dry plate process of photography
1872	Motion picture principle first demonstrated
1873	Maxwell announces electromagnetic theory of light
1874	Triple expansion engine introduced by A.C. Kirk.
1875	Rotary perfecting printing press invented
1876	Alexander Graham Bell invents the telephone.
	Nikolaus Otto invents internal combustion engine
	Edison invents the phonograph
1878	Edison develops bi-polar dynamo
1879	Edison perfects incandescent electric lamp
	Van Siemens introduces first electric street car
1880	Charles Laveran discovers malaria germ
	Karl Eberth discovers typhoid germ
1882	Robert Koch discovers tuberculosis germ
	Edison builds first electric power station
1883	John Carbutt introduced celluloid film for photography
	Diphtheria germ identified and isolated
1884	Koch discovers cholera germ
	Boston-New York long distance telephone line opened.
	Sir Charles Parson invents steam turbine
1885	Pasteur develops innoculation against hydrophobia and milk pasteurization
	Morgenthaler invents linotype
	Lanston invents monotype
	Hertz demonstrates existence of radio waves and discovers photo-electricity.
	First submarines built
1886	Gottlieb Daimler invents internal combustion engine

	Smokeless power and high explosions introduced
1887	Edison invents motion picture machine
	Daimler develops first automobile
1889	Dunlop introduces pneumatic rubber tire
	J. L. Firm invents straight-line press
	Edison perfects motion picture using Eastman film
1893	Walter Scott designs first color press
1894	Sir Hiram Maxim designs heavier than air machine.
	First gasoline automobile built
	S. Kitasato discovers bubonic plague germ
1895	Roentgen discovers X-ray
	Rudolf Diesel invents diesel engine
	Guglielmo Marconi invents wireless telegraphy
	Sigmund Freud introduces psychoanalysis
1896	Anti-typhoid inoculation introduced by Sir A. E. Wright.
	Samuel P. Langley flies first airplane model
	Ludwig Obry invents gyroscope
	Becquerel discovers uranium radiation
1897	French 75 mm gun introduced
1898	Santos-Dumont builds first non-rigid airship
	Curies discover radium
	Count Ferdinand von Zeppelin introduces rigid airship
1900	Planck propounds the quantum theory
1902	Valdemar Poulson invents arc transmitter for wireless telegraphy
1903	Wright Brothers make first successful airplane flight
1904	Rutherford discovers alpha particle
	First sound motion picture introduced
1905	Einstein propounds the quantum particle theory and theory of relativity
1906	Sir Frederick G. Hopkins discovers vitamins
	First dreadnought launched
	Smith & Urban develop first motion picture color photography

1907	Leo de Forest patents three electrode vacuum tube
1908	Rutherford-Geiger counter invented
1909	Admiral Robert E. Peary claimed discovery of North Pole
	Henry Ford builds Model-T exclusively
	Louis Bleriot flies across English Channel
1911	Capt. Roald Amundsen discovers South Pole
	Rutherford propounds the nuclear theory
1913	Bohr develops theory of hydrogen atom
	Typhus brought under control
1915	Fokker synchronizes machine gun fire thru revolving propeller
1916	Sikorsky develops twin-motored plane
1918	Germans develop long-range artillery
1919	Alcock and Brown fly across the Atlantic
	Rutherford transmutes nitrogen into oxygen
1920	Juan de la Cierva designs autogyro
	First radio broadcasting introduced
1924	First circumnavigation of globe by airplanes
1925	First successful talking pictures introduced
1926	Cdr. Richard Byrd and Floyd Bennett fly to North Pole
1927	Lindbergh flies from New York to Paris
	First television signals transmitted
	First trans-Atlantic telephone service opened
1928	Geiger-Muller counter invented
	Capt. Charles Kingsford-Smith makes first trans-Pacific flight
1929	Cdr. Richard Byrd flies to South Pole
	"Graf Zeppelin" circumnavigates globe
1930	Pluto discovered
1931	Auguste Piccard makes first flight into stratosphere
1932	Lawrence invents cyclotron
	Curie-Joliots and Chadwick discover the neutron
1932	Urey discovers hydrogen isotopes

1933	Curie-Joliots discovers artificial radioactivity
1937	Russians fly over North Pole from Moscow to Vancouver
1939	Hahn and Strassmann announce discovery of nuclear fission
1941	Kerst develops betatron
1945	Atomic bomb exploded over Japan
1950	Hydrogen bomb development authorized

Summary of First Century of the Aquarian Age

The first decan of the Aquarian Age is modified by Libran influences; hence, we find wars dominated by revolutionary ideologies—socialism, communism, and fascism. The mind of Humanity is swayed by propaganda—the war of ideas preceded the actual clash of arms. It took a civil war to abolish slavery in the United States, but Women were emancipated by constitutional means in Australia (1894), the United States (1919) and England (1928). The desire for liberty in thought and action resulted in rebellion against restrictions and conservatism, the espousal of new, untried, and so-called progressive movements in art, literature, science, economics, politics, and religion.

Vast material wealth was produced and destroyed through increased technological inventions, materialism outstripped the moral and spiritual growth of mankind, and the powers of the mind word often used to enslave others through propaganda of clashing ideologies. The development of new methods of communication such as the press, the movies, radio and television, while increasing the diffusion of knowledge, is creating a race of morons—people who find it easier to accept ideas formulated by others rather than to think for themselves. Hence the rapid spread of "Me too-ism"—the lunatic fringe in religion, politics, economics, etc.

Aquarius is represented by the Water Bearer pouring a stream of water, symbolizing the servant of humanity who pours out the water of knowledge to quench the thirst of the world. This knowledge, when put to proper use benefits mankind, but when prostituted for personal gain and power, works immeasurable harm. Uranus, the ruler of Aquarius, is the planet of science and invention; hence, the remarkable advances in those fields, especially in electricity, electronics, and aviation. The Aquarian ideals of humanitarianism, altruism, and brotherhood were often betrayed for political purposes.

Those who think that the coming of the Aquarian or any other age, or that the coming of a new Messiah, Redeemer, or world-teacher will change the condition of humanity in the "twinkling of an eye" are doomed to disappointment. The transition from one age to another has invariably been a difficult period for mankind—the karma of past wrongs is not wiped clean in a moment, but must be worked out over eons of time until the purified ego returns to its spiritual source.

Great progress has been made in the past one hundred years, and greater progress will undoubtedly be made in the future centuries of the Aquarian Age. Mankind will not destroy itself with either the atomic or the hydrogen bomb. The true Aquarius decan will undoubtedly usher in the Brotherhood of Man.

Conclusion

This research into the coordination of time and space was undertaken to answer the following questions:

Question 1. Has civilization moved westward more or less in harmony with the westward movement of the equinoctial point?
Answer 1. Yes, the historic sequence has been: Euphratean (Sumer and Babylon), Egyptian, Aegean (Crete, Mycenae, Greece), Roman, Teutonic, French, Iberian (Spain and Portugal), English, American.

Question 2. Has one nation dominated in each constellation age?
Answer 2. No.

Question 3. Has the constellation influence dominated all of civilization or only that portion lying within the constellation boundaries?
Answer 3. While all of civilization has been influence by the predominant constellation urge, the extent to which individual nations have responded has been conditioned by their stage of development and their geographic limitations. For example, the Maritime Nations of Western Europe were the ones to logically benefit from the great navigational urge of the true Piscean decan.

Question 4. From which directions have the attacks on our Western civilization come?
Answer 4. From the north and east with three exceptions. The Euphratean civilization was destroyed by the Assyrians from the north, and they in turn were destroyed by the Persians from the east. The Egyptian civilization was first invaded by the Hyksos from the east, and then followed a thousand year struggle between Egypt and Asia, with the final success resting with the East—Assyria, the Neo-Babylonians, and the Persians. Crete, Mycenae, and Rome were destroyed by the barbarians from the north. The Byzantine Empire was destroyed by the Moslems from the east. French civilization was assaulted by the Germans from the east. The English were assailed by the French and Spaniards and later the Germans, from the east. Today the West is threatened by the Russians from the north and east.

The only exceptions to the general northern and eastern attacks were the Roman Empire, the Moslem Conquest, and the Russian Empire, which spread both east and west. The following abortive attempts were made by the West to conquer the East—Alexander the Great whose empire disintegrated at his death, the Crusades, Napoleon, and Hitler.

Chapter 6

The Solar System

Ancient traditions assert that the earliest forms of worship, as well as time measures, were stellar, then lunar, and lastly solar. As early as 6000 B.C., the Egyptians, according to Lockyer, oriented their temples to the stars—thus they were called stellar temples, of which, the Great Pyramid at Giza was perhaps the greatest. Other temples were oriented to the Sun and were known as solar temples, perhaps the most famous being that of Amen-Ra at Karnak, which was built in such a manner that at sunset of the Summer Solstice—that is, on the longest day of the year—the sunlight entered the temple and penetrated along the axis to the sanctuary. Temples oriented in the manner served as astronomical observatories of very high precision, for by them the length of the year could be determined with the greatest possible accuracy.

King Solomon's temple at Jerusalem was so oriented, that when the Sun's rays on the morning of the Spring Equinox shone into the Holy of Holies, they were reflected to the worshippers, whose backs were to the Sun, by the jewels on the breastplate of the High Priest. In Egypt the unit of time was the solar year, which commenced at the Summer Solstice, coincident with the inundation of the Nile, which was heralded by the heliacal rising of Sirius as far back as 4236 B.C. In Babylonia, the unit of time was the month, since the standard of time was the Moon. Following the Exodus from Egypt, the Israelites abandoned the Egyptian time measure in favor of the Babylonian.

Our stellar system is estimated to contain over 100 billion stars, which are the power-houses and building blocks of the universe. The stars are at such vast distances from us that astronomers have coined new terms of measurements. The nearest star, Proxima Centauri, is 4.3 light years away; that is, a ray of light whose speed is 186,300 miles per second will take 4.3 years to go from this star to the Earth. This is 270,000 times the sun's distance from the Earth. A light year equals 6×10^{12} miles; therefore, Proxima Centauri is 125×10^{12} miles away. A still larger unit is called the PARSEC, which is a combination of the words parallax and second. It is equal to 3.26 light years and is the distance to a point whose parallax is 1: of arc. Sirius, the brightest star to our eyes, is 27 times as bright as our sun, yet Rigel is 10,000 times as bright as our sun, but because it is further away its light does not appear to be as great. Sirius is 9 light years and Rigel 540 light years away.

The ancients grouped the stars to represent mythological beings, animals, birds, fishes, etc. These groupings are called Constellations, but modern astronomers use constellation groupings as convenient means of locating stars. Forty-eight constellations were listed in Ptolemy's *Almagest*, written in 150 A.D., while today there are 88. Davidson believes that the dividing line between constellations of Taurus and Gemini marked the Autumnal Equinox of 4699 B.C. (see chapter 2). The ancient Egyptians used a list of stars rising at ten day intervals to mark off their week of 10 days, whereas the Babylonians used a 28-star list and a week of 7 days. The 7-day week was first introduced into Europe by Constantine.

The zodiac is the band of the celestial sphere, 16° in width, centered on the ecliptic (the apparent path of the Sun). It contains the Sun, Moon, and the planets with the exception of Venus and Pluto. The signs of the zodiac are twelve equal divisions, each 30° long, marked off on the ecliptic eastward of the vernal equinox, and named after the constellations of the zodiac with which they coincided in the time of Hipparchus more than 2,000 years ago. They are: Aries, Taurus, Gemini, Cancer, Leo, Virgo, Libra, Scorpio, Sagittarius, Capricorn, Aquarius and Pisces. Owing to the precession of the equinoxes, the Vernal Equinox has moved about 30° westward and the signs of the zodiac with it. Hence, when the Sun enters the sign of Aries on March 21, it is actually in the constellation of Pisces.

The Sun is an average star, but it is the most important member of our solar system, not only because it contains nearly 99.9% of all the mass of the system, and thus regulates the movements of the other bodies, but also because it is the only self-luminous body, with the exception of comets. All the other members of our solar system shine by light reflected from the Sun, which is also the source of our heat and energy. Scientists admit that recent developments in astronomy and related fields suggest a more intimate relationship between man and his cosmic environment than has heretofore been generally acknowledged.

The Sun is a rotating globe of intensely hot gases (55% hydrogen, 44% helium, and 1% of other elements). Helium was actually discovered in the Sun before it was discovered in the Earth. The temperature is 10,000°F at the surface and millions of degrees at the center. It is 864,000 miles in diameter, or 109 times the diameter of the Earth. Its volume is 1,300,000 times that of the Earth, and its mass 330,000 times that of the Earth. Its axis of rotation is inclined 7.25° to that of the ecliptic, areas near the equator completing one rotation in 25 days, but areas 60° from the equator require 35 days to complete one rotation. This aerial rotation affects the amount of solar radiation reaching the Earth.

Each square foot of the Sun's surface radiates 8,000 hp of energy, or a total for the entire surface of 343×10^{21} K.W. The source of this radiation is believed to be atomic energy. While the Earth only receives solar radiation at the rate of 1/6 hp per square foot of surface at right angles to the radiation, it means the difference between life and death on this planet. The weather bureau records the exact amount of solar energy falling on a horizontal surface at 20 stations by means of pyrheliometers.

The amount and intensity of radiation varies with the growth of sunspots, which are cyclonic masses of electrically charged particles ascending from the interior of the Sun and erupting from its surface for distances of thousands of miles. These electrically charged particles bombard space, including the Earth, travelling with the speed of light. The spots originate at middle latitudes, moving across the face of the Sun and disappearing near the equator. They reach their maximum number at about 11-year intervals, but since the electrical or magnetic polarity reverses as the spots cross the equator, the true cycle may be considered to be 22 years on the average. Sunspots produce profound effects in the electrical state of the Earth's upper atmosphere and thus affect the aurora, the intensity and frequency of magnetic storms (which cause disturbances in communication and electric power lines), the Earth's magnetism, the growth of tree rings, plant life, the weather and economic conditions.

Sunspots more nearly in line with the Earth are more effective in producing terrestrial effects than those occurring on regions of the Sun remote from the equator. The intensity of ultra-violet light from the Sun during the sunspot maxima is from 2 to 2½ times as great as during sunspot minima. Such variations in ultra-violet light are accompanied by changes in the state of the Earth's atmosphere, which produces noticeable biological and physiological effects.

Furthermore, differences in solar radiation occur at different times of day and night, at different seasons, and in different climates. These are due to the changing angle at which the Sun's rays enter the Earth's atmosphere and penetrate through to the Earth from time to time and place to place. Thus, near Boston, Massachusetts (42N, 71W) during December-January, on clear days, nearly twice as much solar energy is received on a south facing vertical surface than on a horizontal surface of the same area. During the spring and fall there is not much difference, while during the summer, the Sun being high in the sky, more sunshine reaches the horizontal surface. At Los Angeles (34N, 118 W) the noon-day altitude of the Sun on June 21 is 79.5°. A give amount of solar energy received over a 12 inch surface. On December 22, when the noon-day altitude of the is 32.5°, the same amount of energy is spread over 22 inches.

More solar energy is received around noon than during the early morning or late afternoon. During December-January, the amount of solar energy received before 8:00 a.m. and after 4:00 p.m. (solar time) is less than 3.2% of the total amount received during the entire day. It is well known that death occurs most frequently in the early morning hours, when man has been cut off from the daily accumulation of solar energy for the longest time.

One theory of the cause of sunspots is that tidal waves are produced on the Sun by the planets, which tends to set the whole solar atmosphere into oscillation. Each planet, in turn, starts its own peculiar tidal oscillation. The composite tidal wave at any moment, therefore, depends upon the positions of the planets in respect to one another and the Sun. H. H. Clayton, the famous authority of world weather, constructed a curve based on the positions of seven planets that shows a striking similarity to the curve of sunspot numbers. He used the mean period of a pair of planets when they were at their maximum distance from the equator. H. Voight, a German investigator, constructed a chart in which the influences of Neptune, Uranus, Saturn and Jupiter were combined to give a close resemblance to the sunspot curve from 1749 to 1928.

The American people have found themselves either in a war or in an economic depression during periods of low sunspot activity at approximately 22-year intervals, with the sole exception of 1889, as the following tabulation proves. The low period chosen is that marking the transition of spots from the Southern Solar Hemisphere to the Northern Solar Hemisphere:

Year	Sunspot Number	Historical Event
1755	9.6	French and Indian War (American phase of Seven Years War)
1775	7.0	American Revolutionary War
1798	4.1	Undeclared naval war with France
1823	1.8	Depression, Holy Alliance and Russian threats result in Monroe Doctrine
1843	10.7	Debt Repudiation Depression
1867	7.3	Post Civil War Reconstruction Depression
1889	6.3	Bloodless Revolution establishes Republic of Brazil
1913	1.4	Depression caused by second Balkan War leading to World War I
1933	5.7	Depression—birth of New Deal—rise of Hitler
1953	?	Depression or World War III?

Sunspot values shown are the extreme lows of the cycle.

Gillette states that weather disturbances are caused by influxes of solar electrons in five ways: (1) By generation of atmospheric currents, (2) by condensation of atmospheric moisture, (3) by increased windiness, (4) by increased evaporation, (5) by reduced influx of solar heat. Only alternate peaks of the sunspot cycle are accompanied by much rainfall in the Northern Hemisphere. The peak date of weather cycles is July 7, at which time a given planet and the Earth are very nearly in heliocentric conjunction in celestial longitude 284°, in which region they cross a stream of electrons flowing out of the Sun and tend to produce an abnormal number of sunspots. Other critical regions when the Earth crosses streams of solar electrons are: longitude 164° on March 6 and longitude 44° on November 8. (It will be noticed that these electronic streams are 120° apart.)

All the members of our solar system, including the Sun, rotate on their axes from west to east, or in a counter-clockwise direction. The orbits of these bodies are ellipses of small eccentricity, with the Sun at a common focus, and are nearly in the same plane, with the exception of Pluto whose orbit is inclined 17° to that of the Earth. Our solar system obeys Newton's Law of Universal Gravitation, which states: "Every particle of matter attracts every other particle with a force that varies directly as the product of their masses and inversely as the square of the distance between them." Our entire Solar System is moving with the Sun approximately in the direction of the bright star Vega at the rate of 12 miles per second, or 40,000 miles per hour.

There are nine known principal planets and several thousand minor ones called asteroids or planetoids. The principal planets in the order of increasing distance from the Sun are: Mercury, Venus, Earth (and Moon), Mars, Jupiter, Saturn, Uranus, Neptune and Pluto. Mercury and Venus being between the Earth and the Sun are called inferior planets, while the others are called superior planets. The Moon is considered to be the Earth's satellite, and similar satellites accompany Mars, Jupiter, Saturn, Uranus, and Neptune.

Bode's Law, discovered in 1772, gives the relative distances of the planets from the Sun interms of astronomical units (93,000,000 miles–the mean distance of the Earth from the Sun). Although Uranus, Neptune, and Pluto, and the asteroids were undiscovered at the time Bode enunciated his Law, the following table shows the relationship to hold true for all but Neptune and Pluto:

	Mer.	Ven.	Earth	Mars	Aster.	Jup.	Sat.	Ura.	Nep.	Plu.
Constant	4	4	4	4	4	4	4	4	4	4
Factor	0	3	6	12	24	48	96	192	384	768
Astro. Unit	0.4	0.7	1.0	1.6	2.8	5.2	10.0	19.6	38.8	77.2
Actual	0.39	0.72	1.0	1.52	2.8	5.2	9.54	19.2	30.0	40.0

The following table gives some of the more important data on the planets:

Planet	Million Miles from Sun	Diameter (Miles)	Sidereal Period	No. of Satellites	Tidal Force at Sun
Mercury	35	3,100	88 d.	0	1.10
Venus	67	7,600	224.75 d.	0	2.11
Earth	93	7,900	365.5 d.	1	1.00
Mars	141.5	4,200	687 d.	2	0.03
Jupiter	483	86,700	12 y.	11	2.17
Saturn	886	72,400	29.5 y.	9	0.11
Uranus	1,782	30,900	84 y.	5	0.02
Neptune	2,793	32,900	165 y.	2	-
Pluto	3,675	3,600	248 y.	0	-

Mercury's orbit is more eccentric and more inclined (7°) to the plane of the ecliptic than any other planet, except Pluto. It is a lifeless sphere with little or no atmosphere and has great extremes of temperature, the bright side reaching 350°C and the dark side 273°C. It is believed to always turn the same side toward the Sun, rotating on its axis once in 38 days; hence, the great temperature range. Its light is simply reflected sunlight, but it only reflects 7% of the light it receives.

Venus is the Earth's nearest neighbor and the brightest object in the sky next to the Sun and Moon. When it is ahead of, or west of the Sun, it rises before the Sun and I called the "morning star." When behind the Sun it is seen in the western sky after sunset and is then called the "evening star". Its surface is always covered by a dense cloud bank that reflects 59% of the sunlight that strikes it, hence we don't even know its period of rotation or what is beneath the clouds that are largely composed of carbon dioxide. Venus is thus truly symbolical of the eternal feminine—veiled in mystery. It is known to have a constant day and night temperature, which is below freezing.

The Earth rotates on its axis daily from west to east, hence the heavenly bodies appear to move from east to west. It furthermore revolves annually around the Sun from west to east in an orbit inclined 23.5° to the plane of its equator. This inclination is the cause of our seasons, and the climatic effects produced thereby have certain definite physiological and biological effects. Thus, according to Huntington of Yale University, the best temperatures for physical activity are 60° to 65°F, and for mental vigor 39° to 54°F. Eminent people tend to be born in greater numbers in January, February, and March, but are conceived in maximum numbers at a temperature some 6 degrees to 8°F lower than that at which the maximum number of conceptions take place among people in general. Petersen, of the University of Illinois, states that more males are conceived in the early part of the year, more females in the relatively stable summer and autumn. Deaths from certain disease reach highs in late winter and early spring.

Since the Earth is slightly flattened at the poles, and bulges at the equator, it is called an oblate spheroid. The difference between its polar and equatorial diameters is 26.9 statute miles or 23.3 nautical miles. The Earth's atmosphere provides the oxygen we breathe, shields us from the destructive ultraviolet rays, prevents a change of temperature of several hundred degrees between day and night, and carries the moisture that prevents the Earth from becoming a desert. The electrified position of our atmosphere, called the ionosphere, makes radio communication possible. It is, however, affects by sunspots that make circuit changes necessary to avoid disruption of communications.

The Moon is the Earth's dead satellite, rotating on its axis from west to east in about 27.33 days. It also revolves around the Earth from west to east in an elliptic path, its nearest approach to the Earth being 221,000 miles and its farthest being 253,000 miles. Its orbit is inclined 5°8' from the plane of the Earth's orbit, and the points of intersection are called Nodes. The Sun must be near a node to produce either a solar or lunar eclipse. The same side is always presented to the Earth. Since it has no atmosphere, the temperature ranges from plus 200°F to -200°F. It exerts a gravitational force on the Earth that causes the tides. Since the Moon and the Earth both revolve around the Sun, the length of one revolution of the Moon with respect to the Sun is 29.5 days (the synodic month). Actually, the Earth and Moon revolve around each other, and it is the center of mass of the two that follows the path regarded as the Earth's orbit.

Scientists of the Bureau of Standards have found that the daily variation of the Earth's magnetism is caused by solar and lunar influences. At the Peruvian experimental station it was found that during the four months centered on the December solstice, the amplitude of the magnetic variation is about 70% greater four days after New and Full Moons than it is four days after the 1st and 3rd Quarters. This is attributed to the direct effect of tidal motions of the atmosphere.

Although moonlight has only 1/300,00 of the intensity of sunlight, Stetson found marked changes from eight years of observations of field-strength determinations of radio waves between Chicago and Boston. The field strength changed from a relatively low value of 75 microvolts a few days after New Moon, to a relatively high value 138 microvolts about two days before Full Moon. This was found to be due to abnormal reflections from the E layer of the ionosphere around the time of Full Moon and reduced reflections at New Moon. The phases of the Moon are based on the angle of elongation of the Moon from the Sun. The elongation is 0 degrees at New Moon, 90° at 1st Quarter, 180° at Full Moon, and 270° at 3rd Quarter.

Petersen finds male deaths to be numerous and female deaths few during Full Moon. Deaths from cardiovascular diseases and tuberculosis occur in greater numbers after Full Moon than before. In general, deaths by suicide show a minimum at Full Moon, but the more slender types apparently commit suicides at New and Full Moons. The menstrual cycle in women is strongly influenced at New and Full Moons. Solar eclipses occur at New Moon, when the Moon is between the Sun and Earth, and lunar eclipses at Full Moon, when the Earth is between the Sun and Moon. Eclipses of a series follow each other at intervals of 18 years 10.33 or 11.33 days, which interval is called a Saros cycle. The ancient Babylonians were expert at predicting eclipses.

Mars is conspicuous for its red color and great brilliance. It has a rarer atmosphere than ours, and the temperature varies from 50°C to -85°C. It has seasons comparable to those of the Earth, but they are nearly twice as long, and the Marian day is slightly longer than the Earth's. Vegetable life on Mars seems possible; but the absence of oxygen seems to bar high forms of animal life. Its two satellites are named Phobos (fear) and Deimos (panic), appropriate companions of the war god, Mars.

According to Bode's Law there should be a planet between Mars and Jupiter. On January 1, 1801, Piazzi, an Italian astronomer at Palermo, discovered a tiny planet 480 miles in diameter, which was named Ceres. More than 1,500 are now known to exist, but only a few are more than 100 miles in diameter and the combined mass of all asteroids probably does not exceed 1/1000 of the Earth's mass. On October 30, 1937, Hermes, which is one mile in diameter, came within 485,000 miles of the Earth. One theory holds that asteroids are the remains of a planet that exploded in space.

Jupiter is the largest of the planets, its mass exceeding that of all the others combined. Because its period of rotation is a little less than 10 hours, its equatorial bulge is comparatively large. Because of its greater distance from the Sun, it is much colder (-140°C) than Earth. The outer part of the planet is believed to have a very dense atmosphere of ammonia and methane with ammonia predominating. Its three outermost satellites revolve from east to west.

Benner quotes Tice in *Elements of Meteorology* (1875) to the effect that "equinoctial disturbances in Jupiter affect the Sun, and through the Sun the solar system. The result upon the Earth and its atmosphere is an enormous increase of electric intensity. Jupiter is the cause of the atmospheric, telluric, and solar perturbations that occur once and in a modified form twice in every one of his orbital revolutions, and the maximum disturbance upon the Earth must occur at or near Jupiter's equinox, and the energy of the equinox of any planet is intensified when that of another occurs at or about the same time." It would appear that Tice anticipated the later findings of Stetson.

Saturn is next in size to Jupiter, but less dense, being the only planet lighter than water. It has a deep cold (-150°C) atmosphere, probably composed of ammonia and methane, with methane predominating. Its equatorial bulge is the most prominent of any planet. Its outermost satellite, Phoebe, revolves from east to west. It has three distinct rings composed of many small particles, similar to meteorites. The outer ring is 170,000 miles in diameter and, each of the rings is 10,000 miles wide and not more than 10 miles in thickness. They are in the plane of Saturn's equator, which is tilted about 28° to the plane of its orbit. Twice during each revolution of Saturn, we see the rings on edge; they appeared thus in 1921 and 1936.

Uranus was the first planet to be discovered with a telescope. It had been observed a number of times before William Herschel in 1781 concluded that it was a planet. It is half as dense as the Earth, and its atmosphere probably consists of ammonia and methane. The orbits of its satellites lie in the plane of the planet's equator, which is almost at right angles to the plane of its orbit around the Sun; therefore, they move almost at right angles to the direction of the planet's motion. If their direction of revolution, which is the same as the direction of rotation of the planet, is considered as direct, the inclination of their orbits to the ecliptic is nearly 98°. If the inclination is considered but 82°, the motion of both planets and satellites is retrograde, or from east to west.

Neptune, discovered in 1843, has a low density, with a deep, cloudy atmosphere and temperature of -222°C. ts satellites revolve from east to west. Since it was first seen only 100 years ago and its sidereal period is 165 years, its complete orbit has not as yet been observed.

Pluto, discovered February 18, 1930, has a more eccentric orbit than the other planets, but since its sidereal period is 248 years, but little of its orbit has been observed. Because of its great distance and small size, we know less about Pluto than any of the other planets.

Comets were once looked upon as harbingers of disaster, but, while still the most spectacular objects in the heavens, they are no longer feared. A great many are known to exist, an average of five being discovered each year, of which two-thirds prove to be return visits of previously known comets. The greatest number ever seen in one year was 14 in 1947. They are very unstable, travel in highly eccentric orbits, inclined at any angle to the ecliptic, and have widely varying periods of revolution. A comet consists of a large nebulous body called the coma, at the center of which is a solid star-like nucleus, sometimes as bright as the brightest stars, and a spectacular, gaseous tail, which may be millions of miles long. Nothing unusual was observed when the Earth passed through the tails of the comets of 1861 and 1910. The most famous comet of this century was Halley's, who concluded that its appearance in 1682 was the return of the comets of 1531 and 1607. He therefore predicted its reappearance in 1759, which prediction was fulfilled. It reappeared in 1835 and 1910, and is due to reappear in 1985. Cowell and Crommelin of the Greenwich Observatory have traced 28 appearances back to 240 B.C. Amongst them were those of 1066 A.D., when William of Normandy conquered England, and that of 1456 A.D., three years after the capture of Constantinople by the Turks.

Meteors or shooting stars are small, solid particles, usually no larger than a tiny bit of gravel, that seem to originate in comets. When they enter our atmosphere, they are heated to incandescence by the friction of the air, most of them being completely consumed before reaching the Earth. The few that survive and strike the Earth are called meteorites. Meteor showers occur when the Earth encounters a swarm of meteors, the most conspicuous being the Perseids or August meteors in the constellation of Perseus, which begin the middle of July, when the Earth is in longitude 284°, and last till the middle of August, reaching their maximum on August 11. The most remarkable shower of the century occurred on the evening of October 9, 1946, in the constellation of Orion.

While seasonal showers of meteors have been attributed to the Earth's crossing hypothetical orbits in which meteoric meteor travels, Gillette believes it more probable that fixed streams of solar electrons bring such matter into our atmosphere. Spiraling solar electrons, attracted by the South Magnetic Pole, account for the great number of meteoric showers in that region, where sunspot numbers are also greater. Electrons omitted around the axis of cyclonic sunspots have the same direction of whirl as that of the sunspot, and consequently have the magnetic polarity of the Sun's atmosphere from whence they escape. Hence, they are magnetically attracted by magnetic foci in the opposite hemisphere of the Earth.

Chapter 7

Time

The foundation of an orderly routine of settled community life, as well as the arts of celestial navigation, surveying, and sidereal geopolitics, is time, which is based upon the apparent angular motions of the heavenly bodies in reference to the Earth—the Sun by day and the Moon and Stars by night. The ancient civilizations measured time by means of (a) gnomon or shadow-pole, which eventually developed into the obelisk or shadow-clock, (b) the pyramids facing the equinoctial sunrise or sunset, (c) the temples with their corridors and portals oriented to greet the rising or setting Sun, or the transit of a guardian star, (d) the great stone circle of Stonehenge, with its sight-line pointing to the rising Sun of the summer solstice.

Let us briefly review the historical development of time measurement. Ancient men saw the Sun rise on one side of his horizon and set on the opposite side. He called the place of sunrise—the eastern horizon, and the place of sunset—the western horizon. He noticed that the shadow of a tree or of a stock or spear stuck upright in the ground varied in length throughout the day. The shadow became shorter as the Sun rose higher in the sky, becoming shortest when the Sun was at its highest point (which he called noon), and then lengthened as the Sun began to descent in the west. The noon shadow was seen to always point to the same place on the horizon, which he called the north point. The opposite point on the horizon he called the south point (see Figure 24). lines connecting the north and south points, and the east and west of the horizon, thus provided the first terrestrial coordinates of position.

The ancient Babylonian priests about 2000 B.C., had devised a system of enumeration called the sexagesima1 system, having 60 for its base, since that figure had the greatest number of exact divisors, namely, 1, 2, 3, 5, 6, 10, 12, 15 and 30. Having learned to make an angle of 60° by inscribing a figure of six equal sides (hexagon) within a circle, the whole circle was considered to contain 360°. The Sun's apparent daily rotation was assumed to take place in a circular path of 360°. One-half of 360° or 180° was assumed to represent the daylight portion of the Sun's path, and the other half the night portion. Since the Sun's noon shadow divided the daylight portion in two, morning and afternoon were each 90° of the circle. This was then divided by 6 into 15° divisions. Thus, 360° divided by 15° = 24 divisions, called hours. Each hour was then subdivided into 60 divisions called minutes, which in turn were subdivided into 60 divisions called seconds.

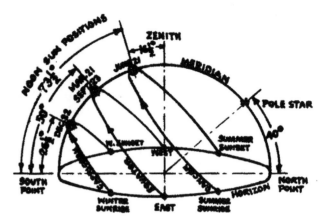

Fig. 24, Seasonal variations of Sun's altitude at Philadelphia, 40N

Thus, was born the shadow-clock or obelisk, which usually took the form of a vertical pillar mounted on a circular stone base that was graduated into equal 150 divisions, representing hours (see Figure 25). But the shadow-clock does not record a constant hour throughout the year because the angle which the shadow of a vertical pole makes with the meridian (azimuth angle) depends upon the Sun's declination, which varies with the seasons. During the progress of the seasons, the places of sunrise and sunset gradually move northward from the date when the noon shadow is longest (the Winter Solstice on December 22) to the time of the Summer Solstice on June 21,

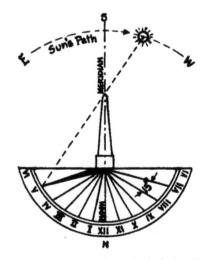

Fig. 25, Shadow clock on obelisk, Sun's shadow shortest at noon; noon shadow shortest June 21, longest December 22

when the noon shadow is shortest. For example: The noon shadow of a 6 ft. pole at 40° north latitude varied from 12 ft. in midwinter to 2 ft. in midsummer. Thereafter, the places of sunrise and sunset begin to shift southward until they again reach the points from which they started. Only at the vernal equinox on March 21 and the autumnal equinox on September 23, are day and night equal throughout the Earth. This is due to the fact that while the Earth rotates on its axis daily in the plane of the equator, it revolves annually around the Sun in an elliptical orbit in the plane of the ecliptic which is inclined 23.5° from the plane of the equator.

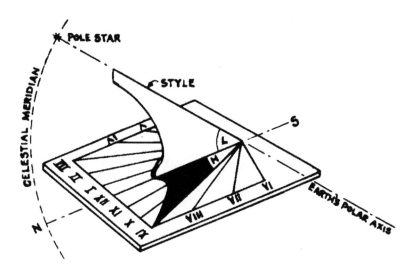

Fig. 26, Sun dial keeps true solar time for latitude; shadow scale formula: tan H = sin lat. tan hour angle

To overcome the limitations of the shadow-clock, the sundial was invented (see Figure 26). In this device the "style" or pointer is oriented to the Pole star, that is, the pointer is set along the north-south meridian, and its edge is inclined at an angle equal to the latitude of the place; the scale is then graduated into divisions or hours, which are constant for all seasons of the year. The formula for graduating the scale is: tan shadow angle, sin Latitude X tan hour angle. For example: Required the shadow angle for 10 a.m. or 2 p.m. in London; tan H = sin 51 degrees tan (2h x 15 degrees) = 0.7771 x 0.5774 = 0.4487 = 24.15°. Since the edge of the sundial pointer is set for the latitude of the place, a sun dial set for London (51N) will not record the correct time in New York (40N). Later modifications of the sundial had an adjustable "style" that made it possible to use the device anywhere, provided that the "style" was correctly set for the latitude of the place of observation. The sundial is the only timepiece that measures time by the motion of the true (or apparent) Sun. When mechanical clocks were invented, they were checked for accuracy against the sundial.

The earliest mechanical timepieces were (a) the sandglass, which is still used to time the boiling of an egg, (b) the clepsydra, or water clock, dating back to the time of Hammurabi of Babylon (ca. 2133-2080 B.C.), (c) the hour candle of the medieval monasteries. By measuring the outflow of sand from a sandglass or the outflow of water from a clepsydra, the priests of Babylon were able to record the time elapsing between one star's transit of the meridian of the observer and the transit of the same meridian by another star. Since the heavens appear to rotate 360° in 24 hours, the angle between two stars that transit the meridian an hour apart, is 15°. Since the Pole star was always seen directly over the spot to which the noon shadow of the shadow-clock point sunset and set on the western horizon at sunrise, divided the night when it was directly over the Pole star, which time was called midnight.

Figure 27 shows how the time of transit of stars or star clusters is measured. When star A in the constellation of Cassiopeia is directly above the Pole star, its hour angle is zero, or its clocktime is 24 (midnight). star B is then approximately 30° east of A, hence it will take about 2 hours for Cassiopeia to transit the meridian. The "pointer" stars in the constellation of Ursa Major (the Big Dipper) are 135° east of star B, hence they will transit the meridian 9 hours later. The star A (caph) is important because it is near the Vernal Equinox, and when it transits, or crosses the local meridian, the local sidereal time is "0" h.

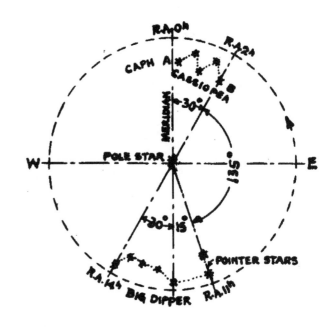

Fig. 27, Measuring star transit time

Figure 28 shows how to tell time by the position of the "pointer" stars in the Big Dipper. The 9 p.m. position for each month is shown, as well as the intermediate hours for July 1. The stars transit the meridian 4 minutes earlier each night, or 2 hours earlier each month. Thus, a star that transits the meridian at 10 p.m. on November 1 will transit at 8 p.m. on December 1. The observer will see the constellations follow each other as the Earth continues on its path around the Sun.

A great circle passing thru the north and south points of the horizon and the zenith of the observer (the point directly overhead), is called the observer's Celestial Meridian and coincides with his Meridian of Longitude. The instant at which any point of the celestial sphere or any heavenly body is on the meridian of the observer, is known as the transit, culmination, or meridian passage of that point. The time elapsing between two successive transits of a heavenly body over the same meridian, is called a day. In the case of the Sun, it is called a solar day; in the case of the Moon a lunar day; in that of a star a sidereal day.

In modern observatories, the time elapsing between two successive upper transits of the Vernal Equinox is called a sidereal day, which begins at sidereal noon, or 12 hours ahead of the civil day, which starts with the Sun on the lower meridian. It is the fundamental unit of time and is ap-

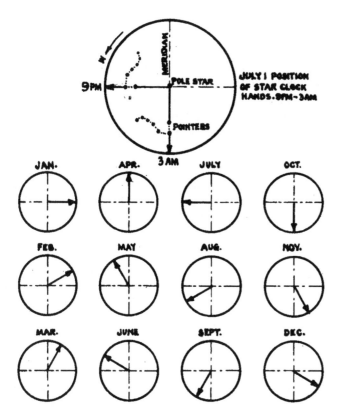

Fig. 28, Telling time by Big Dipper monthly 9:00 p.m. positions

proximately 4 minutes (3m55s.909) shorter than the solar day, because the Earth is not only rotating on its axis, but it is also moving in its orbit around the Sun. The upper branch of the meridian is that portion containing the observer's zenith, while the lower branch is the opposite half. It is noon when the Sun is at upper transit, and midnight when it is at lower transit. The U.S. Naval Observatory determines sidereal time from observations of 18 stars, with an error of only 3/1000 of a second.

The spread of commerce to the Western world, where sunny days were fewer, and the demands of navigators for better navigation tables, spurred the invention of mechanical clocks. The weight-driven clock was invented about A.D. 1000, the spring-driven in the 16th century, and the pendulum clock by Huyghens in 1657. The British government in 1714, offered a prize of 20,000 pounds for a method of determining longitude at sea within an error of 30 miles. (Longitude is obtained by comparing Local Time as obtained from the Sun or stars with standard time at Greenwich.) The prize was won in 1765 by Harrison, a Yorkshire carpenter, for his invention of a chronometer, which on a test voyage to Jamaica in 1761 reached its destination wih an error of only one mile. (This chronometer is still running in London.) LeRoy, of Paris, at the same time invented a better device, which in principle is similar to the modern chronometer, which is set to Greenwich civil time. Thus, the modern method of determining longitude at sea is less than 200 years old. To assist navigators in checking their chronometers, the U.S. Naval Observatory in Washington, D.C., broadcasts time signals from Arlington, Virginia, five minutes before the beginning of each hour.

The length of the solar day varies (1) because the speed of the Earth in its elliptical orbit around the Sun varies, and (2) the projection of the Sun's eastward motion in the ecliptic upon the celestial equator varies. To overcome this objection, an imaginary Mean Sun is used, which moves eastward in the celestial equator at a uniform rate of speed equal to the average rate of speed equal to the average rate of the True Sun in the ecliptic. The interval between two successive lower transits of the Mean Sun is the Mean Solar, or Civil Day, which is approximately 4 minutes longer than the Sidereal day. Time thus measured is called Mean, or Civil Time, and the difference between Apparent (or True Sun) Time and Mean Time at any instant is called the Equation of Time, the values for which are listed in the Nautical Almanac. For example: Sun dial or True Sun Time at noon of February 12 at Greenwich occurs at 12h14m23s G.M.T.

In order to avoid the confusion and inconvenience that arose when each locality kept its own time, Greenwich, England, was designated by international agreement as the starting point for universal time, or Greenwich civil time (GCT), which is measured from the lower branch of the Greenwich Meridian westward to the mean Sun. Prior to 1925, the Astronomical day started at Noon and time was designated as mean time. Since 1925, the Astronomical day was changed to coincide with the civil day which begins at midnight, at which time the Sun is on the opposite side of the Earth, and the time was designated as civil time. The 24 hours of the day are divided into two equal 12 hour periods. That following the lower transit of the meridian is designated a.m. (ante-meridian) and, that following the upper transit is designated p.m. (post-meridian). In astronomy and navigation, the hours are numbered consecutively from "0"h at lower transit to 24h at the next lower transit. Only at the instant of Greenwich noon does the same date prevail throughout the Earth.

In 1884, a majority of the nations of the world agreed to the setting up of 24 standard time zones centered on Greenwich, each 15° wide (or one hour apart). The clocks of a given time zone are set to show the average or mean time of its central meridian. Standard time in the U.S. is divided into four time zones, i.e., Eastern, Central, Mountain and Pacific, which are centered on the following meridians of longitude: 75°, 90°, 105°, 120° W. Since difference in time equals difference in longitude and conversely difference in longitude equals difference in time, time in these zones is therefore 5, 6, 7, and 8 hours, respectively, earlier than Greenwich. At varying dates, Daylight Saving Time has been introduced by various countries.

Local civil time is the true mean or civil time for a given longitude. To change from standard time to local civil time (a) decrease the Standard Time by 4m for every degree the place is west of the standard meridian. For example: 5:00 p.m. Eastern standard time in Baltimore, Maryland (76° W) is 5.00 - .04 = 4:56 p.m. Local Civil Time. (b) Increase the standard time by 4m for every degree the place is east of the Standard meridian. For example: 5:00 p.m. Eastern standard time at New York, New York (74° W) is 5.00 + .04 = 5.04 p.m. local civil time.

In 1934, the *Nautical Almanac* tabulated values (at hourly intervals of GCT) for the Greenwich hour angle (GHA), which is the angular distance measured from the upper branch of the Greenwich meridian westward to the celestial body. In other words, it is the westerly longitude of the Geographical Position of the celestial body at a given instant of GCT. For example, the GHA of the true Sun is the westerly longitude of the true Sun, which is 12 hours greater than the GAT which is measured from the lower branch of the Greenwich meridian. The local hour angle (LHA) is the angular distance measured from the upper branch of the local meridian eastward or westward, whichever is closer, to the celestial body. Thus, the LHA of a body differs from the GHA by the longitude of the observer, or LHA = GHA + Long. E or - Long. W. This factor appeals to navigators, since it eliminates the need for calculating sidereal time.

Sidereal time (ST) measures the rotation of the Earth with respect to the Vernal Equinox, and is identical with the right ascension of the meridian (RAM). Therefore, Greenwich sidereal time (GST) is the Greenwich hour angle of the Vernal Equinox (GHA ♈), or the right ascension of the Greenwich meridian (RAGM). Similarly, (LST) local sidereal time is the local hour angle of the Vernal Equinox (LHA ♈). Therefore, LST = GST + Long. E. or - Long. W. When the Vernal Equinox is on the observer's meridian, the LST is 0h; when the Vernal Equinox has passed the meridian by 15°, the LST is 1h.

Since sidereal time gains nearly 4m on mean or civil time, it is necessary to (a) add 9s.86 for every hour of elapsed civil time after 0h civil time, (b) add 0s.657 for every degree of longitude east of Greenwich. For example: Find the LST (RAM) at Philadelphia, PA (75° W) at 8:00 p.m. EST (20h LCT) on January 1, 1950. From the *1950 American Ephemeris and Nautical Almanac* we get:

(1) GST for 0h GCT on 1/1/50		6h 40m 18s
(2) Add elapsed time to 8:00 p.m.		.20
(3) Add correction for ST of elapsed time		
(20 x 9s.86 = 197s.2 = 3m 17s)		3 17
(4) Add correction for difference in Long. W		
(75° x 0s .657 = 49s.28)		49
		26h 44m 24s
(5) Subtract circle of 24 h		24
(6) Answer: LST (RAM) at Philadelphia		2h 44m 24s = 41°06'

The GHA method using the data in the *Nautical Almanac*, is much simpler, as the following shows. In Philadelphia at 8:00 p.m. on January 1, 1950, the navigator's chronometer would read 1h GCT for January 2, 1950, because Philadelphia is 5 hours west of Greenwich. Taking the GHA for the appropriate GCT and subtracting the longitude, gives the answer within 4/10 second of the previous method.

(1) GHA ♈ (GST) for 1h GCT on 1/2/50   116° 06.1
    (20h LCT on 1/1/50 + 5h W Long.)
(2) Subtract Long. W   75
(3) Answer: LHA ♈ (LST or RAM) at Philadelphia 41° 06.1 = 2h 44m 24s.4s

Problems in time are simplified by using a diagram on the plane of the celestial equator called a time diagram per Figure 29. The observer is supposed to be outside the celestial sphere and looking down at its South Pole. When the Mean Sun is on the lower branch of Greenwich meridian (opposite side of the Earth from Greenwich), the GCT is 0000 hours (midnight). Similarly, when the Apparent (True) Sun is on the lower branch of the Greenwich meridian the Greenwich Apparent Time is likewise 0000 hrs. Since the Sun travels westward 15° per hour, a place which is 15° west of Greenwich would have a local civil time (LCT) one hour less than that of Greenwich, or LCT = 2300 hrs. After the mean Sun has traveled westward for 6 hours, the GCT is 0600 hours and the LCT of the place M is 0500 hours. Thus, LCT = GCT + E Long. or - W Long. Other relationships are shown, which aid in understanding the principles under discussion.

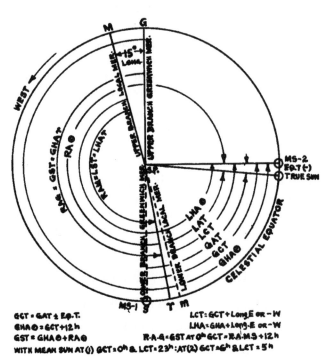

Fig. 29, Time diagram for 0h and 6h, January 1, 1950

Chapter 8

The Calendar and Ancient Chronology

The fundamental units of time in all ages have been the day, the month, and the year; the first being measured by the Earth's rotation on its axis in 24 hours, the second by the Moon's revolution around the Earth in 29 days, 12 hours, 44 minutes, 2.78 seconds (Ed. Note: Strictly speaking this is the time from one lunation to the next. The Moon's revolution in its orbit around the Earth, with reference to the stars, is 27 days, 7 hours, 43 minutes, 11 seconds.), and the third by the Earth's revolution around the Sun in 365 days, 5 hours, 48 minutes, 46 seconds.

The week of 7 days was unknown to the early Egyptians; it is derived from the Chaldeans who named the hours of the day for the 7 heavenly bodies, which they noticed moving in the zodiac, in what was then supposed to be the order of their distances from the Earth: Saturn, Jupiter, Mars, the Sun, Venus, Mercury, the Moon. The day was then named identically with its first hour. Since in naming the 24 hours the 7 names were used three times with 3 names left over, the order of the days became that of Saturday for Saturn, Sunday for the Sun, Monday for the Moon, Tuesday for Mars, Wednesday for Mercury, Thursday for Jupiter, Friday for Venus. Sunday always began at the sunrise following a phase of the Moon, the month beginning with the Sunday following a New Moon.

The ancient Sumerians and Babylonians used the month as their measure of time. It began at the moment the moonsickle (the thin crescent of the Moon) became visible in the western sky immediately after sunset, which occurred every 29th or 30th day, the interval between being called a month. Actually, the Moon becomes "new" when its center point, in its eastward journey around the Earth, passes the center point of the Sun. Because the New Moon is too close to the Sun at that instant, it is usually one or two nights later that the moonsickle is seen. The lunar year contained six months of 29 days and six of 30 days, a total of 354 days.

The lunar year was adjusted to the solar year (365.25) days by periodically intercalating an extra month on decree of the local government. Because the local governments had different ideas, there was a great deal of confusion which persisted to the time of Hammurabi of the 1st Babylonian Dynasty, who unified the calendar throughout his territory. The Assyrians, however, made the adjustment by moving their year one month forward every two or three years. The month whose beginning was nearest the Spring Equinox was called the first month of the year. Tiglath-Pileser I (ca. 1112-1074 B.C.) reformed the Assyrian calendar by introducing the Babylonian intercalation of a month after a certain definite time, in order that the year might always begin in the same month—Nisan—(March-April of our calendar), instead of moving forward from one month to another.

Thus, the reformed Babylonian, Assyrian, and later, the Julian years were of equal length. The Mohammedans, have, however, retained the original lunar calendar, having a year of twelve lunar months, and containing alternately 354 and 355 days, without any intercalated month. Hence, their calendar gains one year in 33 of our calendar years. The Jewish calendar retains the 12 lunar months of alternately 29 and 30 days, with the intercalation of a 29 day month every third year, so that in 19 years there are 7 such leap years. The civil year begins with the New Moon nearest the Autumnal Equinox, in the month of Tishri, which can occur on any day from September 6 to October 6 of the Gregorian calendar. The Moslem and Jewish days both begin at sunset.

The Hindu astronomers in the 7th century B.C. introduced a regular 5-year cycle, in which an extra month was inserted in the second and fifth years. The Greeks used the Metonic cycle, discovered by Meton in 433 B.C., in which an extra month is intercalated seven times in 19 years. This cycle consists of 235 synodic months (from New Moon to New Moon), which is approximately equal to 19 common years of 365.25 days. It is used even today in church calendars to find the date of Easter Sunday.

The Egyptians, according to the religious texts inscribed in the pyramids of the 5th and 6th Dynasties, used a calendar of 12 months of 30 days each (the nearest approximation to the Lunar month of 29.5 days), to which were added 5 supplementary days at the end of the year. The year began on the day the Nile usually reached its crest, which occurred when the Dog Star Sirius (called Sothis) rose just before the Sun (heliacally). But, such a year was one-quarter of a day too short, and the accumulated error became a whole year in 1460 years, called a Sothic Cycle. It was not until 139 A.D. that the Latin author Censorius recorded that the heliacal rising of Sirius coincided with the beginning of the Egyptian calendar year. Modern writers, therefore, assume that the Egyptian calendar came into use in 4236 B.C.

The ancient Romans originally used a calendar of 10 months beginning with March, and totaling 304 days. January and February were added later, making 12 months averaging 29.5 days, or a total of 354 days, an extra month being occasionally intercalated. But the length of the year was arbitrarily from time to time by the priesthood, the magistrates, or the ruler. Hence, by the time of Julius Caesar, the confusion had become intolerable, and the Vernal Equinox was falling in December instead of March. So Caesar called in the Alexandrian astronomer Sosigenes to revise the calendar. In 45 B.C. it was decreed that the year should consist of 365 days, with the addition of an extra day every fourth year. The year was to begin on the 1st of January, which in 45 B.C. was the day the New Moon followed the Winter Solstice. The preceding year (46 B.C.) was made 445 days long to bring the Vernal Equinox forward to March 25, and it is still known as the "Las Year of Confusion." Caesar's revised calendar was called the Julian calendar. The ancient Chinese also used a calendar that has 3 years of 365 days and one year of 366 days, which is traditionally believed to date back to 2697 B.C.

But, the Julian year of 365½ days is too long by 11 minutes and 14 seconds, since the actual Tropical Year (year of the seasons is 365 days, 5 hours, 48 minutes, 46 seconds). Thus, from 325 A.D. to 1582 A.D. the accumulated error amounted to 10 days, and the Vernal Equinox occurred on March 11, 1582 instead of on March 21 as it did at the time of the Council of Nicaea 325 A.D. At that time the Church adopted the following rule for fixing the date of Easter Sunday: "The first Sunday after the 14th day of the Moon (nearly Full Moon) on or immediately following the day of the Vernal Equinox." Pope Gregory XIII, on the advice of the Jesuit astronomer, Clavius, ordered that the day after October 4, 1582, be called the 15th instead of the 5th. The rule for leap years was changed to the following: "All years whose date number is divisible by four without a remainder are leap years, unless they are century years that are not divisible by 400, such as 1700, 1800, 1900, 2100."

The Gregorian calendar was adopted by: Italy, France, Spain, Portugal and Poland in 1582; Holland, Flanders, and most of the German Roman Catholic States in 1583; Denmark, and the German and Dutch Protestant states in 1700; the British Dominions in 1752 (the day following September 2 was called September 14 and 1752 began January 1st instead of March 25—a relic of the time of Caesar when the Vernal Equinox occurred on that date—thus the year 1751 lost the months of January and February, and the first 24 days of March); Sweden 1753; Switzerland 1812; Alaska 1867; Japan 1872; China 1912; Bulgaria 1915; Turkey 1917; Russia 1918; Yugoslavia and Romania 1919; Greece 1923. The Gregorian calendar is good for 3000 years.

The French scholar, Joseph Scaliger, in 1582 A.D. devised the Julian Epoch named after his father, as a universal harmonizer of the different systems of chronological reckoning then in use. This is based on a period of 7980 years, which is the least common multiple of three cycles that were much used in Roman chronology, namely, the Roman Indiction, the solar cycle, and the lunar cycle. The current Julian period began January 1, 4713 B.C. on which date all three cycles coincided. In this system, which is widely used by astronomers, each year is a Julian year of 365¼ days. An astronomical event such as the solar eclipse of August 8, 1896 is expressed as

J.E. tear 6609, 222nd day, or Julian Day 2,413,781. *The American Ephemeris* and *Nautical Almanac* contain a table from which the Julian day number of any date is readily obtained. Thus, January 1, 1950 of the Gregorian calendar is Julian day 2, 433,283, and it begins at Greenwich mean noon. The year 1950 corresponds to the year 6663 of the Julian period, and January 1, 1950 of the Julian calendar corresponds to January 14, 1950 of the Gregorian calendar.

The Julian year is tied in with other chronological eras as follows:

Year	Era	Julian Calendar Date
7459	Byzantine (5508 B.C.)	September 1, 1950
5711	Jewish (3761 B.C. creation of the world	August 29, 1950
2703	Founding of Rome (753 B.C.)	January 1, 1950
2699	Nabonassar (747 B.C.)	April 23, 1950
2610	Japanese	December 18, 1949
2262	Selucidea (312 B.C.)	September 1 or October 1, 1950
1667	Diocletian (284 A.D.)	August 29, 1950
1370	Mohammedan (622 A.D. Flight of Mohammed	September 29, 1950

The Christian Era began with the date of the birth of Jesus as calculated in the 6th century by Dionysius Exiguus, but subsequent investigations based on the computed date of an eclipse that occurred at the time of King Herod's death indicate that Jesus was born in 4 B.C. (Ed. Note: There is other strong evidence based upon historical and astronomical evidence that His birth occurred in 7 B.C.)

The Sumerians dated their years after the most important event that had happened during the previous year, i.e., "the first year after such and such an event." The Babylonians adopted this system from the Sumerians, but with the arrival of the Kassites they began to date their years from the beginning of the King's reign. Thus, Ling Lists were formed, and then Dynasty Lists. The Assyrians, Greeks and Romans dated their years after a high official called in Assyrian limmu or eponym, who was chosen each year by drawing lots. The Assyrian king usually performed his limmuship during the first full year of his reign, so the number of limmu between his limmuship and that of his successor gave the exact number of his regal years. Hence, the reign of an Assyrian king was reckoned from his first full year on the throne until the first full year of the next king. In other nations, the reign included the "Accession" year.

The Assyrians made lists of their eponyms, giving the names of the limmu, their official capacity in the Assyrian state, and a brief note concerning the chief event of the year. From numerous baked clay tablets, which have been excavated from the ruins of the library of Assurbanipal (668-626 B.C.) at Nineveh, 22,000 of which are now in the British Museum, the distinguished Oriental scholar Sir Henry C. Rawlinson, in 1867, discovered the Assyrian Canon, a chronology from about 909 B.C. to 680 B.C. From this and other tablets, A. Ungnad in 1938 compiled a continuous list of limmu called the Eponym Chronicles for the years 890 to 649 B.C. Additional valuable data is provided in the Great List of Assur which goes back beyond 1200 B.C.

One of the most important uses of astronomy to the historian is that by it the exact dates of certain historical events may be calculated and thus the chronology of the ancients put on a firm foundation. This is done by calculating the dates of eclipses. As far back as the 23rd century B.C. the astrologers of Sumer and Akkad observed a Lunar Eclipse, which Karl Schoch calculated occurred on March 8, 2283 B.C. However, such observations were spasmodic, and it was not until the reign of the Babylonian king Nabonassar (747-734 B.C.) that a continuous record began to be made in 747 B.C. which continued for over 360 years. The only long continuous modern record is that of the meridian observations at Greenwich, England, which began in 1750, only 200 years ago. The Office of Naval Research is using the "electronic brain" to calculate the orbits of the navigational planets over the past 400 years on order to provide navigators and astronomers with more accurate tables.

Not long before 500 B.C., Naburimannu, a Chaldean astronomer used the recorded observations of the previous 250 years to compile tables of the motions of the Sun and the Moon, from which he calculated the daily, monthly and yearly revolutions of these bodies, as well as solar and lunar eclipses, etc. Without the aid of telescopic observations, he calculated the length of the year as 365 days, 6 hours, 15 minutes and 41 seconds, or 26 minutes and 55 seconds longer than the modern value. About a century later, another Chaldean astronomer, Kidinnu, made a similar set of tables of greatly increased accuracy, from which he discovered the precession of the equinoxes. This was definitely established by Hipparchus about 125 B.C., but the true explanation had to wait for Sir Isaac Newton (1642-1727). The more accurate tables which will result from the research project of the U.S. Nautical Almanac Office, should prove of incalculable value.

Rawlinson, in the May 18, 1867 Athenaeum quotes from and Assyrian tablet that had been described in the British Museum Report for 1854 as follows: "In the eighteenth year before the accession of Tiglath-Pileser there is a notice to the following effect: 'In the month Sivan an eclipse of the Sun took place, and to mark the great importance of the event a line is drawn across the tablet, although no interruption takes place in the official order of Eponymes. Here then we have notice of a solar eclipse, which was visible at Nineveh, and occurred within 90 days of the Vernal Equinox (taking that as the normal commencement of the year) and which we may presume to have been total from the prominence given to the record, and these are the conditions that during a century before and after the Era of Nabonassar are alone fulfilled by the eclipse that took place on June 15, 763.

This is believed to be the eclipse that the prophet Amos foretold in 787 B.C. in Amos VIII, 9: "And it shall come to pass in that day, saith the Lord God, that I will cause the Sun to go down at noon, and I will darken the Earth in the clear day."

Prof. Van der Meer of the University of Amsterdam gives the translation as follows: "During the Epnymate of Pur-Sagale, Governor of Guzana, revolt in the City of Assur. In Simanu there was an eclipse of the Sun." Van der Meer states that the June 15, 763 B.C. date of this eclipse "is the sheet anchor upon which depends not only Assyrian chronology but also that of the whole of Western Asia." Pur-Sagale was the eighth limmu after Assurdan III (772-755 B.C. With the aid of the Khorsabad King List, discovered in the palace of Sargon III by the Oriental Institute of the University of Chicago in 1932-33, Van der Meer has carried Assyrian chronology backward to the reign of Samsi-Adad I (1726-1694 B.C. plus 20 years) without a break, except for an estimated length of 20 years for the reigns of Assurabi and Assurnadinahhe I about 1428 B.C. The chronology was then carried forward to the reign of Assurbanipol (668-626 B.C.), the last great ruler of Assyria, whose capital, Nineveh, was destroyed by the Medes and Babylonians in 612 B.C.

The fixed point in Babylonian history that can be linked up with a fixed year in Assyrian history is given in the Babylonian Chronicle I, 29-32, which reads: "In the fifth year Shalmaneser died in the month Tebet. Shalmaneser had exercised the kingship of Akkad and Assyria for five years. In the month Tebet, on the twelfth day, Sargon seated himself on the throne of Assur. In Nisan, Meredach-baladan seated himself on the throne of Babylon." Since Shalmaneser ruled in both Assyria and Babylon for five years, from 727 to 722 B.C., Sargon of Assyria and Meredach-baladan of Babylon each ascended their respective thrones in 722 B.C., their first regal years being 721 B.C., because Tebet is the tenth month of the year (actually January in the Julian Calendar) and the reign doesn't begin until the following Nisan, which is the first month of 721. The chronology is carried backwards to the reign of King Messannipadda, founder of the Ist Dynasty of Ur (ca 2472-2433 B.C.) and then forward to the destruction of Babylon under Nabonnidu (ca. 553-537 B.C.) by Cyrus the Great in the summer of 537 B.C.

In 1933, Andre Parrot of the Musee du Louvre recovered 20, 000 clay tablets from the palace of Mari, an ancient city on the Middle Euphrates River. Some of the tablets contain diplomatic correspondence between the last king of Mari, Zimri-Lim, Hammurabi of Babylon, and Samsi-Adad I of Assyria, indicating that they were contemporaries. Since we have previously seen that the latter ruled about 1748-1716 B.C., the correspondence indicates that Hammurabi ruled about 1728-1686 B.C., Van der Meer gives the dates as: Samsi-Adad I $1726 + X$ to $1694 + X$ and Hamurabi $1712 + X$ to $1670 X$ where = approx. 20 years. This brings the date of Mannurabi forward about 300 years from the date previously given by historians, via: 2123-2020 B.C. in Hammerton-Barnes *Illustrated World History* and about 1950 in Langer's *An Encyclopedia of World History*.

There are a number of historical correlations between the chronology of Assyria and Palestine, such as: (1) The Battle of Karkar, which took place in 853 B.C. during the sixth year of

Shalmaneser III (859-852 B.C.). (Ahab was slain by the Syrians the year after the Battle of Karkar): (2) Shalmaneser furthermore records on the Bull Inscription of Shalmaneser that in his eighteenth year he received tribute from King Johu of Israel (841-814 B.C.) who ascended the throne after slaying King Ahaziah of Judah and King Joram of Israel in 841, 90 years after the death of King Solomon. (3) The Kingdom of Israel was destroyed when Samaria fell to the Assyrians in 722 B.C. in the ninth year of Hoshea (730-722 B.C.) which was the third year of Hezekiah of Judah 727-698 B.C.) The Kingdom of Judah was destroyed when Nebuchadnezzar of Babylon (605-561 B.C.) captured Jerusalem in 586 B.C. in the eleventh year of Zedekiah (597-587 B.C.). In Israel, the king's reign included his accession year, while in Judah it did not. Jerusalem fell in 587 of the Jewish year, which began in Tishri (October-November), or 586 of the Babylonian year, which began the following Nisan (March-April).

The following are some of the historical correlations between the chronology of Egypt and Western Asia. (1) The Tell el-Amarna tablets show that Amenophis IV or Ikhnaton (1375-1358 B.C.) was a contemporary of Ashurbellit of Assuria (1362-1327 B.C.). (2) Ramesses II (1292-1225) B.C.) signed a peace treaty with Hattushilish the Hittite king in 1271 B.C. The latter wrote a letter to Kadasman-Enlil of Babylon (1266-1256 B.C.) congratulating him on his accession to the throne in 1266 B.C. (3) In the 5th year of Rehoboam of Judah 931-914 B.C.) Shishak I of Egypt (945-924 B.C.) captured Jerusalem in his twentieth year (925 B.C.) (4) Necho of Egypt (609-593 B.C.) was defeated by Nebuchadnezzar of Babylon (605-561 B.C.) at the Battle of Carchemish (605 B.C.). (5) Psamtik III of Egypt was defeated by Cambyses II (529-522 B.C.) at the Battle of Pelusium (525 B.C.) and Egypt became a Persian province.

Chapter 9

Eclipses

Eclipses of the luminaries—the Sun and Moon—are the earliest astronomical events recorded. They were regarded in ancient times with fear and awe, being sometimes looked upon as harbingers of disaster, unless the proper propitiatory rites were performed by the priests. Thus, when the Chinese astronomers royal Hi and Ho got drunk during the reign of Chung-Kan, fourth emperor of the Hsia Dynasty, and failed to predict the solar eclipse of October 22, 2136 B.C., they were beheaded as punishment and to appease the gods. Confucius, in one of his canonical books—the *Chun-Tsen*—gives a short review of 36 solar eclipses observed in China between February 20, 720 B.C. and July 22, 495 B.C.

The Chinese, Hindus, and some American Indian tribes believed that a monster was trying to swallow the luminaries during an eclipse, and could only be driven away by making loud noises with drums, trumpets, horns, bells, fire-crackers, etc. The Incas believed that a solar eclipse was proof that they had offended the Sun, who thereby expressed his anger. The Chaldaeans, according to Canon Rawlinson, studied eclipses for the purpose of making predictions regarding "storms, tempests, failing or abundant crops, war, famine, and the like, for Syria, Babylonia, and Susiana; but they could venture no prophecies with respect to other neighboring lands, as Persia, Media, Armenia." Riccioulus states, "that there were treaties written to show against what regions the malevolent effects of any particular eclipse was aimed, and the writers affirmed that the effects of an eclipse of the Sun continued as many years as the eclipse lasted hours, and that of the Moon as many months."

Breasted records that "as far back as the 23rd century B.C. in the days of the kings of Sumer and Akkad, the astrologers observed an eclipse of the Moon, which has now been calculated by a modern astronomer (Karl Schoch) to have occurred on March 8, 2283 B.C." While there is no doubt that the Babylonian priests observed and calculated eclipses, which were carefully noted and recorded, most of the records were destroyed in 747 B.C. by Nabonassar, who sought thereby to have exact chronology begin with his reign. A catalogue of eclipses recorded on baked clay tablets dated from 2233 B.C. is said to have been discovered by Alexander the Great and sent to Aristotle. Unfortunately, all of the tablets have been lost, but Ptolemy, who is believed to have had access to this list of others dating from the time of Nabonassar, preserved a record in his *Almagest* of six eclipses dating from 721 B.C.

In 1653 A.D., J. B. Ricciolus published a catalogue of eclipses observed from 772 B.C. to A.D. 1647 with tables to A.D. 1700. In 1757 James Ferguson reprinted all of the eclipses recorded by Nicolas Struyck of Amsterdam in 1740, dating from 721 B.C. to A.D. 1485, and extended the list to A.D. 1800. The French astronomer Pingre in 1783 published a very important and comprehensive catalogue of solar and lunar eclipses from 1000 B.C. to A.D. 2000. The most complete and comprehensive catalogue of eclipses is that of Theodore Von Oppolzer of Vienna, published in 1887, which gives approximate calculations for the visibility of 8,000 solar and 5,200 lunar eclipses from November 10, 1207 B.C. Julian calendar to A.D. November 17, 2161 Gregorian calendar.

Kepler, using the Rudolphine Tables, was perhaps the first astronomer to develop a method of computing eclipses with some approach to scientific form. His method was improved by Flamsteed, but modern calculations are based on the work of Hansen and Bessel, and the data is published annually in the Nautical Almanacs of the leading nations. For all practical purposes, the paths of eclipses may be calculated from Newcomb's "Tables of Solar Eclipses from 700 B.C. to 2300 A.D." published in *Astronomical Papers of the American Ephemeris and Nautical Almanac 1882*.

What are eclipses and when do they occur? Astronomically, an eclipse is the darkening of a heavenly body, especially the Sun or Moon, by the shadow of another body. All members of our solar system cast conical shadows (which are normally invisible) as they travel around the Sun. An eclipse of the Sun occurs when the Moon's shadow (232,100 miles long) falls on the Earth. An eclipse of the Moon occurs when the Earth's shadow (859,000 miles long) falls on the Moon. See Figure 30. When the Moon's shadow is long

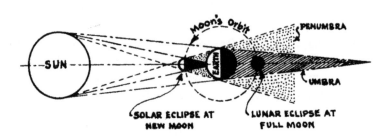

Fig. 30, Solar and lunar eclipses

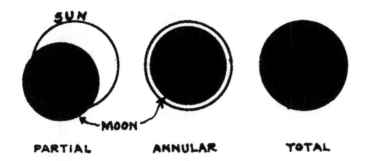

Fig. 31, Types of solar eclipses

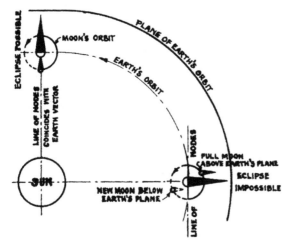

Fig. 32, Eclipse seasons, Moon's orbit inclined 5°8′ to plane of Earth's orbit preclude monthly eclipses

enough to touch the Earth, we have a total eclipse, but when the shadow fails to reach us, the Moon will not quite cover up the face of the Sun, leaving a ring of light around the Moon's edge, hence the name Annular Eclipse. Observers within 2,000 miles either side of the Moon's shadow will see a partial eclipse (see Figure 31). If the orbit of the Moon were in the same plane as the earth, there would be an eclipse of the Sun at every New Moon, and an eclipse of the Moon at every Full Moon. But the Moon's orbit is inclined 5°8′ to the Earth's orbit (ecliptic), therefore, the Moon's shadow usually passes above or below the earth at the time of New Moon, while the Moon at Full Moon usually passes above or below the earth's shadow. It is only when the New or Full Moon occurs near a Node that an eclipse occurs (see Figure 32). The Nodes are the points where the Moon's orbit intersects the ecliptic, the one where the Moon crosses the ecliptic from south to north being called the ascending or North Node (Dragon's Head), and the other, the descending or South Node (Dragon's Tail).

The Nodes of all planets, including the Moon, regress or move westward along the ecliptic because of the perturbations caused by the mutual interaction between the planets. Thus, in Figure 33, if the Moon's motion were unperturbed, it would cross the ecliptic at N; but the Sun's perturbation action causes the Moon

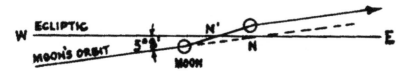

Fig. 33, Westward regression of Moon's Nodes

to move more nearly at right angles to the ecliptic and crosses at N' to the west, and thus the Moon's Node regresses. After Node passage, the movement is again toward the ecliptic and the inclination is restored after a slight temporary disturbance. The westward movement of the nodes is completed in 18.6 years.

The time during which an eclipse can take place is about five weeks and is called an eclipse season, of which there are two—in March and September 1951, and about 20 days earlier each succeeding year because of the westward travel of the nodes. If at the time of Full Moon, the Earth's shadow is more than 12°15' from a node, a lunar eclipse is impossible, but if less the 9°30', a lunar eclipse is inevitable. In the case of solar eclipses the major ecliptic limit is 18°30' and the minor 15°21' (see Figure 34). The least possible number of eclipses in a year is two, both central eclipses of the Sun, and the largest possible number is seven, five solar and two lunar.

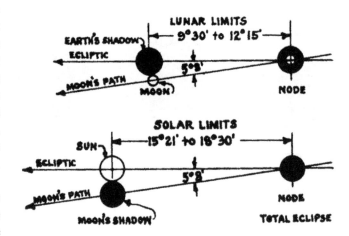

Fig. 34, Solar and lunar ecliptic limits

In 1950 there were four eclipses, two of the Sun and two of the Moon, but only one, a total eclipse of the Moon on September 26, 1950, was visible from Washington, D.C. In 1951, two annular eclipses of the Sun will be visible from the United States, the first on March 7 being partially visible from Washington, with 22% of the Sun's diameter covered by the Moon at the middle of the eclipse. The second will be partially visible from Washington on September 1, with 92% of the Sun's diameter covered by the Moon at the middle of the eclipse.

Since the shadow of the Earth, where the Moon passes through it, is 5,700 miles in diameter and the Moon travels 2,000 miles an hour, it may remain partly in the shadow for 4 hours, and more than 1 hour if it passes centrally through the shadow for a total eclipse. On the other hand, the Moon's shadow falling directly on the Earth is only 167 miles, and its speed being 2,000 miles an hour while the Earth at the equator is travelling in the same direction at 1,000 miles an hour, the net speed of the shadow is 1,000 miles an hour. Hence, a total solar eclipse never lasts more than 7½ minutes, and considerably loss if the eclipse occurs near sunrise or sunset, or at higher altitudes where the shadow's speed may reach 4,000-5,000 miles an hour.

While lunar eclipses may be seen from any place on the night side of the Earth, a total Solar

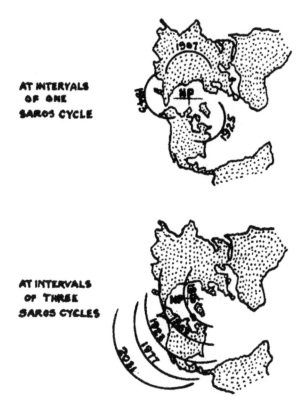

Fig. 35, Geographical movement of eclipses

eclipse can only be seen by observers within the relatively small area (umbra) upon which the Moon's shadow falls. Thus, no total eclipses have been visible from London between A.D. 878 and 1715, and only two since then—the next one being scheduled for August 11, 1999. In general, only one total solar eclipse may be seen in 360 years from any give place on Earth. Many people may, however, witness partial eclipses of the Sun, since the penumbra or partial shadow of the Moon is 4,000 miles in diameter where it reaches the Earth. Eclipses of the Sun always begin on the west side of the Sun, while eclipses of the Moon begin on the east side of the Moon, because the Sun's movement in the ecliptic is only apparent, while that of Moon is real. The frequency of eclipses is in the ratio of two lunar to three solar, and annual eclipses (when the Moon is farthest from the Earth) are more frequent than total, in the ratio of 3 to 2.

Eclipses are roughly predictable by the Saros cycle, known since the time of Thales in the 7th century B.C., and probably discovered by the greatest astronomers of antiquity, the Chaldaeans. It is a period of 6,585.33 days, or 18 years 11.33 days (10.33 days if there are 5 leap years), at the conclusion of which, the centers of the Sun and Moon return very nearly to their relative positions at the beginning of the cycle. Thus, the annular solar eclipse of March 18, 1950 and the total solar eclipse of September 12, 1950 are returns of the famous ones of 1842, 1860, 1878, 1896, 1914 and 1932. However, the one-third day of the Saros cycle causes each return to occur 120 degrees further west, thus requiring 54 years 1 month to return to the same longitude (see Figure 6). Solar eclipses run through 70 and 71 Saroses totaling 1,260 years during which their tracks move from one pole across the face of the earth to the opposite pole. Thus, the eclipses of 1865, 1883, 1901, 1919, 1937, 1955, 1973 move northward, while those of 1850, 1868, 1886, 1904, 1922, 1940, 1958 move southward. In about 6,000 years a new series of eclipses begins, but at the opposite pole from whence the previous series began. Lunar eclipses run through 48 or 49 Saroses totaling 865 years.

Regression of the Nodes produces a very noticeable change in the diurnal path of the Moon in different parts of the 18.6 year cycle, for when the ascending Node coincides with the Vernal Equinox, the Moon's path lies outside the ecliptic and its maximum declination is 23°27' + 5°08' = 28°37'. This occurred in 1931 and 1950. But when the ascending Node reaches the Autumnal Equinox about 9 years later, the maximum declination is only 23°27' - 5°08' = 18°19' (see Figure 36). Thus, in 1950 the maximum monthly range in declination was from +28°37' to -28° 27', a total of 57°14', whereas 9 years earlier it was from +18°19' to -18°19', a total of 36°38'.

Total Solar Eclipses and U.S. Events

Date	Duration	Region	Historic Event Precedes
June 16, 1806	4.30 min.	New York, New England	War of 1812
Nov. 30, 1834	2 min.	Arkansas, Missouri, Alabama Georgia	Panic of 1837
July 18, 1860	3 min.	Washington and Labrador	Civil War 1861
Aug. 7, 1869	2.45 min.	Iowa, Illinois, Kentucky, North Carolina	Panic of 1873
July 29, 1878	2.30 min.	Wyoming, Colorado, Texas	-
Jan. 11, 1880	32 sec.	California	Garfield Assassination
Jan. 1, 1889	2.15 min.	California, Montana	Panic of 1895
May 28, 1900	2 min.	Texas to Virginia	McKinley Assassination
June 8, 1918	2.4 min.	Washington to Florida	Armistice
Sept. 10, 1923	3.6 min.	California to Mexico	Death of Harding
Jan. 24, 1925	2.4 min.	New York and New England	Coolidge Election
Aug. 31, 1932	1.5 min.	New England	Roosevelt Election
July 9, 1945	-	Idaho, Montana, North Dakota	Use of "A" bomb
June 30, 1954	-	Nebraska, South Dakota, Minnesota, Wisconsin	

Total eclipses of the Sun are the greatest of all astronomical spectacles, and even modern astronomers are struck with awe when the light in the sky takes on an eerie aspect and queer shadows dance across the ground as the Moon's shadow moves across the face of the Earth at tremendous speed. Little wonder then that primitive peoples viewed such an event with fear and trembling. Several classic instances have been recorded in history, perhaps the most famous being the total eclipse of May 28, 585 B.C., which so frightened the Medes and Lydians, who were engaged in battle, that they immediately broke off the engagement and made peace with each other. Herodotus credits Thales of Miletus with having predicted the eclipse. Xenephon records that

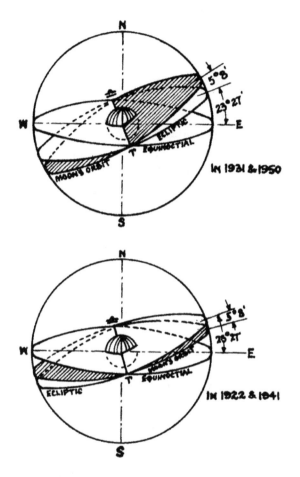

Fig. 36, Changing declinations of Moon's orbit

the Persians under Cyrus captured the City of Larissa from the Medes who were frightened by the solar eclipse of May 19, 557 B.C., which occurred during the siege of that city.

The ancient astrologers believed that "Eclipses of the luminaries always signify a change of affairs." Thus, the defeat of the Persians under Xerxes at the Battle of Salamis is believed to have been foreshadowed by the Solar eclipse of February 17, 478 B.C. Similarly the eclipse of August 3, 431 B.C. at Athens, in the first year of the Peloponnesian War is believed to have foretold the defeat of Athens by Sparta in 404 B.C. Agathocles of Syracuse interpreted the eclipse of 310 B.C. as favorable to his cause when he was blocked by the Carthaginian fleet, because, as Justin states: "Some change was certainly signified, either to Carthage which was in such a flourishing condition, or to them (the Greeks) whose affairs were in a very ruinous state." Agathocles escaped from the Carthaginians and proceeded to ravage their home territory in Africa.

A great annular eclipse of the Sun occurred when Julius Caesar crossed the Rubicon on January 6, 49 B.C. to conquer Italy and lay the foundations of the Roman Empire. The eclipse of 840 A.D. in Bavaria is said to have frightened to death Emperor Louis, whereupon his three sons proceeded to fight each other over the succession, which was finally settled in 843 A.D. by dividing Charlemagne's Holy Roman Empire into France, Germany, and Italy.

Historic Solar Eclipses

Date   *Historic Event*
Nov. 24, 29   Crucifixion of Jesus
June 18, 410   Alaric, King of the Visigoths captured Rome
Feb. 24, 453   Attila the Hun died while ravaging Italy
Aug. 30, 1030   Olaf, King of Norway, killed at Battle of Stiklastad
Aug. 2, 1133   Presaged death of Henry I of England in A.D. 1135
Sept. 4, 1187   Saladin captured Jerusalem in October
June 28, 1451   War between Mohawks and Senecas averted
March 16, 1485   Richard III, king of England, killed at Battle of Bosworth Field
Jan. 24, 1544   2 years before death of Luther; 3 years before deaths of Francis I of France and henry VIII of England
May 3, 1715   Death of Louis XIV of France; Jacobite uprising in England
Oct. 9, 1847   Preceded Year of Revolutions (1848)
Aug. 14, 1914   Defeat of Russians by the Germans

Historic Lunar Eclipses

Jan. 29, 1136 B.C.   First recorded lunar eclipse during Chou Dynasty in China
Mar. 19, 721 B.C.   Meredach-baladan became king of Babylon
Aug. 27, 413 B.C.   Preceded destruction of Athenian fleet and army at Syracuse
Sept. 30, 331 B.C.   Foreshadowed defeat of Darius by Alexander at Arbela, Oct. 1, 331 B.C.
June 21, 168 B.C.   Defeat of Macedonians by Romans at Battle of Pydna
Mar. 13, 4 B.C.   Death of Herod
Mar. 1, 1504 A.D.   Columbus' prediction saves his crew from starvation by Indians
July 4, 1917 A.D.   Lawrence of Arabia captured Turkish fort under cover of darkness
Aug. 14, 1943 A.D.   Axis forces escaped from Sicily to Italy during darkness

Historians frequently use the dates of eclipses to fix the dates of historic events. Thus, the solar eclipse of June 15, 763 B.C., at Nineveh, is the "sheet anchor upon which the chronology of western Asia and Egypt is based." The birth of Jesus was determined from the lunar eclipse of March 13, 4 B.C., which occurred the night before Herod died. (Ed. Note: While this is an eclipse accepted in timing Herod's death, research with which the author apparently is unacquainted, strongly indicates that Jesus was born December 23, 7 B.C. N.S.).

From the time of Eratosthenes, Hipparchus, Pliny, and Ptolemy, until the middle of the 19th century, when the modern chronometer was invented, lunar eclipses were used to determine longitudes on land and at sea. Pliny states that at the time of the Battle of Arbela (331 B.C.), the

Moon was eclipsed at the second hour of the night, when at the same time it was rising in Sicily. Ptolemy also cites this eclipse as giving the distance between Carthage and Arbela. Much greater accuracy was attained by the Arabs in their calculations of longitude, an some of their figures were passed on to the western world in astronomical tables during the 12th and 13th centuries.

The mariner who possessed an almanac giving the local time at which an eclipse would occur at a given place, could, therefore, obtain his longitude by recording the local time of its occurrence at the place where he then was. Thus, if the almanac gave the Greenwich local time of a lunar eclipse as 5:58 p.m., and the mariner saw the Moon's disc entering the Earth's shadow cone at Aden at 8:58 p.m., he would know that he was 3 hours (8:58-5:58 = 3:00) east of Greenwich and that his longitude was 45° E. The ships of Columbus, Vespucci, and Magellan had to put into port to take a bearing in longitude. Thus, Columbus sought a port of anchor at Haiti for observing at rest the conjunction of the Sun and Moon on January 13, 1493. Amerigo Vespucci found the difference in longitude between Venezuela and Cadiz by observation of a Lunar eclipse in 1499 A.D.

Scientists study:

(1) the spectrum of the eclipsed Moon to obtain data on the constitution of our own atmosphere;
(2) the heat radiated during the different phases of a lunar eclipse
(3) stars occulted during a lunar eclipse for data on parallax and the Moon's size.

Solar eclipses are studied:

(1) to furnish better tables of the motions of the Sun and Moon;
(2) to obtain photometric measurements of light intensity;
(3) to search for intra-mercurial planets;
(4) to obtain data on solar prominences, corona, the polarization of light, the spectra of the Sun's lower atmosphere, changes in Earth potential and the vertical potential gradient of our atmosphere.

Chapter 10

Evolution of the Planisphere and House Division

A planisphere is a representation of the celestial sphere on a flat surface. It history dates back to the days of the ancient Babylonian and Egyptian astronomers who used certain groups of stars at 10-day intervals (called "decan" stars) to tell time at night and as navigational aids. Records of these stars have been found in Egyptian sarcophagi lids and the ceilings of tombs of the 6th Dynasty (ca. 2050 B.C.). Although no complete Babylonian planisphere has been preserved, Brown (1900) reconstructed from three fragments of a Euphratean planisphere found by Geo. Smith in the Palace of Sennacherib (705-681 B.C.), and now in the British Museum, a 36-constellation planisphere, of which he says, "Our Planisphere thus takes us back by implication to a period prior to 2540 B.C., and when the Sun was in Taurus at the vernal equinox."

Eisler (1946) describes the reconstruction of an ancient Babylonian planisphere from the broken fragments of a circular clay disc found in the library of King Assurbanipal (668-626 B.C.) and now in the British Museum. Of this planisphere, he writes: "All the 32 stars selected (the other 4 were planets), rose in the 3rd Millenium B.C., at the time of Saragon I of Akkad, in such points of the horizon as to enable the navigator of a ship or the leader of a caravan in the featureless desert to establish an astral compass-card such as Arabic skippers in the Red Sea and the Indian Ocean still used in the early 19th century. Our modern nautical almanacs list the "mean places of the Ten Day Stars" and the "apparent places of Circumpolar Stars" for the benefit of present day navigators.

It is possible that both the Brown and Eisler reconstructions are based on the same broken fragments of the original Babylonian planisphere. Brown's reconstruction shows the 36 constellations in three concentric circles, divided into 12 months, giving a northern, zodiacal, and southern sonstellation for each month. Eisler's reconstruction shows 'Three Stars for Each' of the twelve months arranged in three concentric circles, with a number indicating the time of rising, culmination, or setting of the particular star. Neither the Babylonian nor the Egyptian decan-stars have any relation to the division of each zodiacal sign into three 10-degree decanates, which apparently was a later development of the Greeks.

The earliest surviving Greek planisphere, that of Timochares (328-283 B.C.), shows alpha Draconis as the Pole Star, with quartering circles running thru Spica and Sirius, from which Eisler deduces it to have been copied from a Babylonian planisphere at about 2200 B.C. Similarly, the Egyptian planisphere of Denderah, now in the Louvre in Paris, although reconstructed during the reign of the Roman Emperor Tiberius (A.D. 17) and showing the heavens divided into eight segments, is believed to represent the heavens of about 1700 B.C. Eisler writes, "A comparison of the most characteristic constellations of Denderah with their Babylonian counterparts shows the closest correspondence down to minute details. So it is evident that this sky-map, too, is based on some very old Babylonian, or rather Assyrian, planisphere."

Flammarian, the French astronomer, states that, "Chiron's sphere, the most ancient sphere known, was constructed about the epoch of the Trojan War, 1300 B.C."

According to Cicero, the Greek philosopher Thales of Miletus (636-546 B.C.) made the first known model of the globe, upon the surface of which the astronomer Eudoxus of Kindos (403-350 B.C.) later traced the positions of certain groups of stars. Diogenes Laertios credits Anaximander of Miletus (611-546 B.C.) with drawing the first map of the Earth and construction the first celestial sphere. The Greek poet Aratos (ca. 270 B.C.) versified the constellations shown on Eudoxus' globe, which Brown believes represented the heavens as viewed from Babylon at the Vernal Equinox of 2084 B.C., based upon the calculations of the British astronomer Richard A. Proctor. Brown credits the Euphrates Valley as being the birthplace of the zodiacal constellations saying "The names of the Signs of the Zodiac show that the present arrangement was adopted when the Sun still entered Taurus at the vernal equinox, i.e., between 4698 and 2540 B.C."

The ancient Egyptians, from the time of Menes, ruler of the Southern Kingdom of Upper Egypt, who conquered the Northern Kingdom of Lower Egypt, faced south when worshipping their gods. This direction was called upward or forward, while the north, which lay at their backs, was called the lower or hinder region. The east was at their left and the west at their right. The original home of these southern conquerors of Egypt is believed to have been near the equator, for Lepsius states, "The solstices were always considered as in the horizon, and the Vernal Equinox as up in the sky." This view of the heavens is that of the right sphere (Figure 37), in which the poles lie in the horizon, (Ascendant), reaching their zenith in the Midheaven (MC) and disappearing below the western point of the horizon (Descendant). The Egyptians would therefore represent a plan view of the horizon with its quartering divisions as in Figure 38.

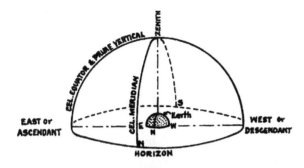

Fig. 37, Right sphere at equator, observer facing south

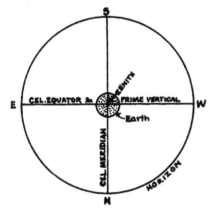

Fig. 38, Right sphere in plane of horizon shwoing quartering of the heavens

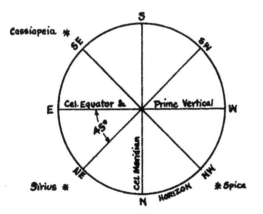

Fig. 39, Earliest compass card of equatorial regions showing eight-fold division of the heavens

Furthermore, the majority of stars in northern latitudes cross the meridian of the observer south of the Zenith, hence sailors speak of the transit of a star as its "Southing." Since the only place on Earth where the Sun always rises due east and sets due west is at the equator, on all days except two, the days of the Vernal and Autumnal Equinoxes. Thus, observers in northern latitudes (north of the Tropic of Cancer) face due south when they face the Sun at noon, at which time it crosses the observer's meridian. Hence, the earliest compass-car of the ancient navigator showed the horizon divided into eight parts, as in Figure 39. In the 3rd millennium B.C., when alpha Draconis was the Pole Star, the four quarters of the heavens formed by the equinoctial and celestial circles were halved by two great circles, one passing through Spica in the northwest to Cassiopeia in the southeast, and the other from Sirius in the northeast to the southwest.

The Greeks sought to connect the phenomena of the heavens with the events of mankind. Hence, they used both the eight-fold and twelve-fold divisions of the celestial sphere as the basis for two systems of mundane house divisions. The Michigan Astrological Papyrus, written in the 2nd century A.D., describes a twelve-house system counted clockwise, and an eight-house system counted counterclockwise. Although the eight-house system has long since been discarded, there are numerous methods of constructing the twelve-house system of which the most prominent are the following:

1. The prime vertical system of Albategnius, Muhammed hen Djabir (890-929 A.D.), which is very similar to the Campanus system, introduced by Joseph and Mathew Campanus in the 13th century, which is based on the tri-section of the Prime vertical, which, at the Equator coincides with the Celestial Equator, as shown in Figure 40. Cyril Fagan, president of the Irish Astrological Society, is the most forceful champion of this system today, and it is gaining wider acceptance in Britain and on the Continent.

2. The equator of the Regiomontanus system, introduced by Johannes Regiomontanus (1436-1476 A.D.), is based on the tri-section of the celestial equator, as shown in Figure 40. This system was widely used on the Continent prior to the introduction of the Placidean system.

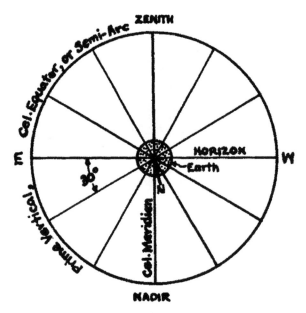

Fig. 40, 3 systems of house division at the equator showing identical intersections of the ecliptic

3. The Semi-arc or Placidean system, introduced by Placidus de Titus in 1603 A.D., is based on the tri-section of the semi-diurnal arcs, which at the equator coincides with the prime vertical and celestial equator and is of equal lengths, as shown in Figure 40. The idea of dividing the Diurnal and nocturnal arcs into equal parts is probably derived from the ancient Egyptian method of timekeeping, which counted 6 hours from sunrise to noon and 6 hours from noon to sunset. Such a shadowclock bearing the name Thetmes III (1479-1447 B.C.) is now in the Berlin Museum. The Greeks adopted this type of clock nearly a thousand years later, and through them the 12-hour day reached Europe. At other latitudes, the equal division of unequal arcs is open to question. This is the system most widely used in America today.

In all three systems, the cusps of the houses are the points where the trisecting circles cut the ecliptic circle, and at the equator they are identical for all three systems. Some writers, however, regard the points on the ecliptic midway between the trisecting circles as the cusps of the Houses in the Campanus system. Although the cusps of the first and tenth houses are identical in all three systems, the cusps of the intermediate houses vary with the latitude. For example, the 1941 *American Astrology Ephemeris* gives the following cuspal values for 40° north:

House System	X	XI	XII	I	II	III
			Sidereal Time, 0h			
Campanus	00♈00	29♈53	10♊19	18♋28	15♌45	07♍27
Regiomontanus	00♈00	08♉33	18♊55	18♋28	10♌49	02♍31
Placidus	00♈00	60♉13	14♊55	18♋28	08♌15	01♍07
			Sidereal Time, 6h			
Campanus	00♋00	26♋58	28♌56	00♎00	01♏04	01♐03
Regiomontanus	00♋00	05♌51	05♍14	00♎00	24♎46	24♏09
Placidus	00♋00	03♌25	03♍50	00♎00	26♎10	26♏35

Tucker (1936), in his Zenith system of equal house division, advocates a return to the method of the Chaldaeans, which consisted of the equal division of the ecliptic by six great circles intersecting in the pole of the ecliptic, see Figure 41. In this system, each house contains exactly 30°, which is not true of the previous systems described. The cuspal degrees for London (51°32′ N) in the Placidean and Zenith systems are as follows:

House System	X	XI	XII	I	II	III
			Sidereal Time, 0h			
Placidus	00♈	09♉	22♊	26♋36	12♌	03♍
Zenith	26♈	26♉	26♊	26♋	26♌	26♍
			Sidereal Time, 6h			
Placidus	00♋	06♌	06♍	0♎00	24♎	24♏
Zenith	00♋	00♌	00♍	00♎	00♏	00♐

Summarizing, we may conclude that the systems of Campanus, Regiomontanus, and Placidus show the least divergence from each other in equatorial regions, the divergence becoming greater with increase in latitude. The Zenith system of equal house division appears to be valid only in regions near the ecliptic; for then the poles of the ecliptic will lie in or near the plane of the horizon. Figure 42 shows that for the latitude of London, the poles of the ecliptic are 23°02′ from the plane of the horizon, in the Zenith system of house division. In actual practice, astrologers seem to disregard the inapplicability of a particular system of house division to particular Latitude by using wide "orbs." From a mathematical standpoint, the Campanus system seems to be the least open to criticism.

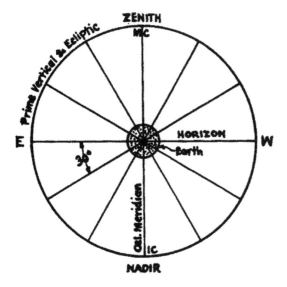

Fig. 41, Zenith system at ecliptic latitude (23.5o); prime vertical and ecliptic coincide; ecliptic poles in plane of horizon

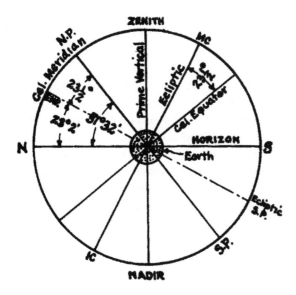

Fig. 42, Zenith system at London; prime vertical and ecliptic do not coincide; ecliptic poles not in plane of horizon

Bibliography

Ageton, Arthur A., Cdr., *Manual of Celestial Navigation*
Alter and Cleminshaw, *Pictorial Astronomy*
Bailey, George H., *Globe-Trotting Round the Horoscope*
Bailey, George H., *The Framework of the Horoscope*
Baker, Robert H., *Astronomy*
Benham, H.E., *Aerial Navigation*
Botsford and Robinson, *Hellenic History*
Bowditch, Nathaniel, *American Practical Navigator*
Breasted, J. H., *Ancient Times*
Brodeur, Arthur G., *The Pageant of Civilization*
Brown, R., Jr., *Primitive Constellations*
Chambers, G.F., *Story of the Eclipses*
Cheiro, *World Predictions*
Councel, Paul, *Cosmic Causation in Geophysics*
Councel, Paul, *Your Stars of Destiny*
Cournos, J., *A Book of Prophecy*
Dakin, E.F., *Mrs. Eddy*
Davidson, D., *The Great Pyramid*
DeVore, Nicholas, *Encyclopedia of Astrology*
Dubberstien and Pletz, *The Glories of Ancient History*
Duncan, John C., *Astronomy*
Durant, Will, *The Story of Civilization*
Dutton, Benjamin, Cdr., *Navigation and Nautical Astronomy*
Eberle and Weems, *Learning to Navigate*
Eisler, R., *Royal Art of Astrology*
Engberg, R.M., *The Dawn of Civilization*
Esslement, J.E., *Bah u'llah and the New Era*
Fagan, Cyril, *The Fundamentals of House Division*
Fagan, Cyril, *The Incidents and Accidents of Astrology*
Fagan, Cyril, *Zodiacs Old and New*
Finnegan, J., *Light from the Ancient Past*
Flammarion and Gore, *Popular Astronomy*
Forman, H.J., *The Story of Prophecy*

Gleadow, R., *The Michigan Astrological Papyrus*
Hall, Manly P., *Astrological Key Words*
Hammerton and Barnes, *The Illustrated World History*
Harding, Arthur H., *Astronomy*
Hayes, Baldwin and Cole, *History of Europe*
Heindel, Max, *The Rosicrucian Cosmo-conception*
Hogben, Lancelot, *Science for the Citizen*
Hogben, Lancelot, *Mathematics for the Millions*
Jensen, L.J., *Astro-Economic Interpretation*
Kraum, Ralph, ed., *Astrological Americana*
Kuhn, A.B., *Theosophy*
Langer, William L., *An Encyclopedia of World History*
Lockyer, Norman J., *The Dawn of Astronomy*
Massey, Gerald, *A Book of Beginnings*
Newcomb, Simon, *Popular Astronomy*
Rawlinson, George, *Five Great Monarchies*
Sepharial, *Geodetic Equivalents*
Shields, B.A., Lcdr., *Meteorology and Air Navigation*
Sidgwick, J.B., *Introductory Astronomy*
Skilling and Richardson, *Astronomy*
Spencer, Katherine Q., *The Zodiac Looks Westward*
Todd, Mabel Loomis, *Total Eclipses of the Sun*
Tucker, W.J., *The Principles of Scientific Astrology*
U.S. Hydrographic Office, *Aircraft Navigation Manual*
U.S. Hydrographic Office, *American Nautical Almanac*
U.S. Hydrographic Office, *Tables of Computed Altitude and Azimuth*
Van der Meer, P., *The Ancient Chronology of Western Asia and Egypt*
Young, C.A., *General Astronomy*

Lightning Source UK Ltd.
Milton Keynes UK
UKHW030659060919
349233UK00008B/1078/P